LASER
GROUP
A. JAVAN

See page ?7
fr f

RESONANCE RADIATION
AND
EXCITED ATOMS

by

ALLAN C. G. MITCHELL, Ph.D.

Professor of Physics
Indiana University

and

MARK W. ZEMANSKY, Ph.D.

Professor Emeritus of Physics
The City College of the
City University of New York

CAMBRIDGE
AT THE UNIVERSITY PRESS
1971

PUBLISHED BY

THE SYNDICS OF THE CAMBRIDGE UNIVERSITY PRESS

Bentley House, 200 Euston Road, London, N.W.1
American Branch: 32 East 57th Street, New York 22, N.Y.

Library of Congress Catalog Card No.: A62-5588

ISBN: 0 521 08433 4

First printed 1934
Reprinted 1961
1971

PRINTED IN THE UNITED STATES OF AMERICA

PREFACE TO THE
SECOND IMPRESSION

In the twenty-seven years since this book was published the interest of physicists in it has changed greatly. At that time, physicists had come to the limit of the then existing techniques in the field and the book was more useful to astronomers who were interested in line shapes existing in the spectra of stars. Within the last few years new fields and new techniques have developed for which the information in this book is of interest. At present these fields are: (1) Those having to do with the determination of spins, magnetic moments and hyperfine structure separations by the methods of "optical pumping" and "optical double resonance"; (2) Optical MASERS; and (3) nuclear resonance absorption of gamma rays.

In view of these developments and the fact that the original edition was out of print, the authors were asked to consider making a new edition. After consultation with the publishers, it was decided to reprint the original edition. This procedure has the advantage of speed and cost. Aside from leaving out a description of the new experiments mentioned above, which can be found by going through the literature, the "reprint method" has the disadvantage that the mathematical approach of twenty-seven years ago may seem somewhat foreign and naïve to the modern reader. The difference in the two treatments is purely one of semantics, however, and the reader can easily translate from the one language into the other. One thing that will be of interest to the present-day reader is that the experiments on the effect of a high-frequency alternating field on the polarization of resonance radiation, described in Chapter V and referred to as the experiments of Fermi and Rasetti and of Breit and Ellett, are the forerunners of the method now known as the method of nuclear magnetic resonance.

A.C.G.M.
M.W.Z.

BLOOMINGTON, INDIANA
NEW YORK, N.Y.
4 *July* 1961

PREFACE TO THE FIRST IMPRESSION

SINCE the discovery of resonance radiation by R. W. Wood in the early part of the present century, a considerable amount of work, both experimental and theoretical, has been done in this field. With the exception of a few articles in the various handbooks of physics, each summarizing a small part of the subject, no comprehensive account of the theories, experiments and interpretations connected with resonance radiation has up to the present existed. As a result, conflicts in notation, experimental method and evaluation of results have arisen which have tended to impede progress. It is the purpose of this book to remove these conflicts, wherever possible, by presenting the theories connected with resonance radiation in an orderly manner with a systematic notation, and by adopting a unified point of view, compatible with modern quantum theory, in discussing and interpreting the experiments that have been performed. Wherever possible, a historical summary has been given, but on the whole, this book has not been written from the historical point of view, but rather, from a critical one. Many of the topics which are treated have their roots so deeply entrenched in classical physics that a historical survey was found both impracticable and unnecessary. Instead, special attention has been paid to the principles and limitations of various methods of studying resonance radiation, and the existing discrepancies and outstanding problems yet to be solved have been critically discussed. Mathematical theory has been introduced into the text wherever it was pertinent, but in cases where a mathematical treatment might be too cumbersome in the text, it has been relegated to an appendix. The bibliographies at the end of each chapter contain references to the most important papers published in the field, but no attempt has been made to list all of the early papers, inasmuch as these may be found in an excellent bibliography at the end of "Fluorescenz und

Phosphorescenz" by P. Pringsheim, published by J. Springer, Berlin.

It is a pleasure to acknowledge our indebtedness to Prof. R. Ladenburg of Princeton University, and to Prof. G. Breit of New York University, for many helpful criticisms. Thanks are due also to the Bartol Research Foundation of the Franklin Institute and to the Physics Department of New York University (University Heights) for stenographic assistance.

We are grateful to the *Physical Review*, the *Philosophical Magazine*, the *Annalen der Physik*, the *Zeitschrift für Physik* and the *Ergebnisse der Exakten Naturwissenschaften* for permission to use certain figures in this book. Under special arrangement with the publishing firm of J. Springer, we acknowledge that the following figures have been taken from the *Zeitschrift für Physik*: Figs. 6, 15, 18, 19, 27, 28, 32, 33, 39, 40, 42, 45, 46, 52 and 53; and from the *Ergebnisse der Exakten Naturwissenschaften* the following: Figs. 71 and 72.

A.C.G.M.
M.W.Z.

NEW YORK CITY
6 *September* 1933

PUBLISHER'S NOTE

Resonance Radiation and Excited Atoms was originally issued in the Cambridge Series of Physical Chemistry, edited by Professor E. K. Rideal.

CONTENTS

Chapter I

INTRODUCTION

Chapter II

PHYSICAL AND CHEMICAL EFFECTS CONNECTED WITH RESONANCE RADIATION

Chapter III

ABSORPTION LINES AND MEASURE-MENTS OF THE LIFETIME OF THE RESONANCE STATE

Chapter IV

COLLISION PROCESSES INVOLVING EXCITED ATOMS

Chapter V

THE POLARIZATION OF RESONANCE RADIATION

APPENDIX

INTRODUCTION

1. GENERAL REMARKS

THE study of "resonance radiation" and related problems is a powerful means of obtaining information concerning the interaction of light and matter. In most cases conditions are so simple that it is possible to get an insight into the behaviour of individual atoms, molecules and light quanta, and thereby to form a foundation on which to build the complicated structure of physics and chemistry. Since "resonance radiation" is, as the name implies, a form of light, and since it is intimately connected with atoms and their structure, it will be necessary to give a brief description* of the present theories of light, atomic and molecular structure, and the "spectra" exhibited by atoms and molecules, before proceeding to the phenomena connected with "resonance radiation".

Until the beginning of the present century all the known properties of light could be explained by the classical electro-magnetic wave theory developed by Maxwell and later brought to perfection by H. A. Lorentz. On this theory light was considered as vibrations in the "ether", the plane of the vibrations being perpendicular to the direction of propagation of the light. This theory was able to explain in a beautiful way the phenomena of polarization, reflection, refraction and diffraction of light waves. Light is known to be "coloured", white light being a superposition of a number of colours, and the colour is intimately connected with the above wave picture. Thus, a given colour is usually defined by its wavelength λ, the distance between crests of the ether waves connected with it, measured (for the visible region at least) in

---

* Only a cursory description of atomic structure can be given in a book of this scope. It is hoped that what is said will be enough to enable the reader to understand what is to follow in the book. A number of books on spectra, atomic structure and atomic dynamics are at hand and the reader will be referred to these where supplementary study is desired.

Ångström units (1 Å. = 10⁻⁸ cm.). The visible region extends from 3800 Å. (violet) to 7700 Å. (red), the near ultra-violet from 2000 Å. to 3800 Å. and the infra-red beyond 7700 Å.

In viewing light emitted from an incandescent solid through a prism spectroscope the colours are spread out into a "spectrum" from violet to red. If the intensity varies gradually from one colour to another, the spectrum is said to be continuous. The intensity distribution as a function of wave-length for light from a hot solid is a function of the temperature of the body (heat radiation). On the other hand, if the source is an electrically excited gas or an electric arc, bright lines are seen on a dark background. This type of spectrum is known as a "line spectrum". Sometimes the lines are very close together and have a "fluted" structure, in which case the spectrum is known as a "band spectrum". If, however, the source is an incandescent gas, like the sun, it is possible to observe dark lines against the bright background of a continuous spectrum. The lines are known as absorption lines and the spectrum as an absorption spectrum. By far the most important types of spectra are "line" and "band" spectra. The former is the radiation given off by atoms and the latter that due to molecules. In the present work we shall be more interested in the line spectra of atoms, so that a brief discussion of this phenomenon will not be out of place here.

2. INTRODUCTION TO LINE SPECTRA

2a. CHARACTERISTICS OF LINE SPECTRA.* In the early part of the nineteenth century Fraunhofer discovered that the sun's spectrum contained a great many dark lines. Later, Bunsen and Kirchhoff discovered that certain elements, when heated in a flame, emitted a spectrum consisting of groups of bright lines. They found at once that the groups of lines obtained were characteristic of the element in the flame. They also were able to identify certain of the emission lines of the flame with absorption lines in the sun. Later developments

* See L. Pauling and S. Goudsmit, *The Structure of Line Spectra*, McGraw Hill Book Company, Inc.

showed that a greater number of lines could be produced by exciting the atom to emission in an electrical discharge than in a flame, the complicated groups of lines that appeared being characteristic of the emitting element.

Most atoms show a rather complicated spectrum, certain of them, however, notably hydrogen and the alkalis, exhibit simpler spectra whose wave-lengths are characterized by series relationships. Hydrogen, the simplest element, exhibits several different series of lines, each lying in a different spectral region. The frequency $\left(\nu \sec^{-1} = \dfrac{c}{\lambda}\right)$, or the analogous quantity, the wave number $\left(\tilde{\nu} \, cm^{-1} = \dfrac{1}{\lambda}\right)$, of any line in these series is given by

$$\tilde{\nu} = R\left(\frac{1}{n^2} - \frac{1}{m^2}\right) \qquad \ldots\ldots(1).$$

In this formula R is a constant, known as the Rydberg constant $(R = 109,677 \text{ cm}^{-1})$, and is the same for all series, the numbers n and m being integers. In a given series n is fixed and m may range from $n+1$ to ∞. The number n is 1 for the Lyman (ultra-violet) series, 2 for the Balmer (visible), and 3 for the Paschen (infra-red) series, and so on. It is seen at once that the frequency of any line in any series is given by the difference of two terms $\dfrac{R}{n^2}$ and $\dfrac{R}{m^2}$ (Ritz Combination Principle), the two quantities being called term values.

As is well known, Bohr was the first to give a theoretical derivation of formula (1). On the assumption that the electron of the hydrogen atom can move in only certain of the classically allowed orbits characterized by quantum numbers n, l and energy E_{nl}, and that when an electron jumps from one orbit of energy E_{nl} to another of energy $E_{n'l'}$ light of frequency

$$\nu = \frac{E_{nl} - E_{n'l'}}{h} \qquad \ldots\ldots(2)$$

is emitted, he derived formula (1) for $\tilde{\nu}$ with the correct value of R. The quantum number l characterizes the angular momentum of the electron in its orbit, and may take values $0, 1, 2, \ldots n-1$. The orbital angular momentum of the atom is $\frac{lh}{2\pi}$, where h is Planck's constant. The quantum number n may take all values from 1 to ∞. The derivation further shows that the term values of Ritz (in cm.$^{-1}$), when multiplied by hc, are actually the energy states given by Bohr. Furthermore, Bohr showed by means of his Correspondence Principle that, whereas the energy states depend only on the principal quantum number n, only those lines appear in which the angular momentum quantum number l changes by ± 1. This is known as the Selection Rule for l.

From Eq. (1) it will be seen that the wave-number separation between two lines becomes less and less as $m \to \infty$. When m is very large the lines are said to approach the series limit. At the series limit the electron is so far removed from the atom that the Coulomb attraction between it and the nucleus is exceedingly small. In this condition the atom is said to be ionized.

The alkali atoms, on the other hand, exhibit a series of doublets instead of the single lines of the hydrogen series. Lumping the two components of the doublets as one, term values can again be found whose differences will give the wave numbers of the lines in question. There are certain quantitative differences between the alkali and hydrogen spectra. An alkali atom consists of a valence electron and a number of inner shells of non-radiating electrons. Due to the electrical screening effect of the internal electrons on the charge of the nucleus, and the occasional penetration of the valence electron into this atom core, states with a given value of n and different values of l may have quite different energies. In hydrogen, on the other hand, the energy differences between states of a given n and different l's are quite small, indeed zero if relativity corrections are neglected. It is therefore necessary in the classification of complicated spectra to designate the state by both quantum numbers n and l. In spectroscopic practice it is customary to designate states as S, P, D, F, etc., where the letters correspond to the values of l as follows:

$$
\begin{array}{cccc}
\text{S} & \text{P} & \text{D} & \text{F} & \cdots \\
l=0 & 1 & 2 & 3 & \cdots
\end{array}
$$

To designate the state further it is customary to indicate the total quantum number of the state by placing it before the letter designating the value of l, thus 1 S, 6 P, 7 S, etc. indicate states whose quantum numbers are, respectively:

$$n=1, l=0; \quad n=6, l=1; \quad n=7, l=0; \quad \text{etc.}$$

Furthermore, the frequencies of the lines emitted in various series are designated by

$\nu = nS - mP$ Principal Series, n is fixed, $m = n, n+1, \ldots \infty$.

$\nu = nP - mD$ 1st Subordinate or Diffuse Series,

$$m = n, n+1, \ldots \infty.$$

$\nu = nP - mS$ 2nd Subordinate or Sharp Series,

$$m = n, n+1, \ldots \infty.$$

This schematic representation shows that the frequency of a line is given by the differences of two terms or energy states, the lower energy state always being given first, the higher second, the order of the terms being independent of whether the line is an emission or absorption line.*

The doublet, or fine structure, of the lines was first correctly described by Uhlenbeck and Goudsmit† on the hypothesis of the spinning electron. Their theory postulates that an electron spins while revolving in its orbit somewhat as the planets spin as they revolve about the sun. The angular momentum due to the spin is $\dfrac{sh}{2\pi}$, where $s = \pm 1/2$. The spin angular momentum $\mathbf{s}$ therefore adds itself vectorially to the orbital momentum $\mathbf{l}$ to give a total angular momentum $\mathbf{j}$:

$$\mathbf{j} = \mathbf{l} + \mathbf{s}.$$

Since a spinning electron is a rotating electric charge it is equivalent to a little magnet. There will, therefore, be a contribution to the energy of a given state due to the interaction of the electron spin (spin magnet) with the magnetic field (equivalent to a current) of the electron revolving about

---

* This scheme seems to be the most generally accepted one and will be used in this book. Some authors prefer to designate emission lines by $mP \rightarrow nS$ and absorption lines by $nS \rightarrow mP$.

† G. E. Uhlenbeck and S. Goudsmit, *Naturwiss.* **47**, 953 (1925); *Nature*, **117**, 264 (1926).

the nucleus. This energy is different, depending on whether **s** is parallel or antiparallel to **l**. If one assumes that the resultant spin for all the electrons forming the atom core is zero, i.e. that the spins of the electrons in the core neutralize each other in pairs, there remains only the spin of the valence electron, in the case of the alkalis. With the exception of S states, which can be shown to be single, all states are actually double. There are, therefore, two states for every value of l (except $l = 0$) with total angular momenta

$$j = l + 1/2, \qquad j = l - 1/2.$$

We can label our states in the following way:

l \\ j	1/2	3/2	5/2	7/2
0	$^2S_{1/2}$			
1	$^2P_{1/2}$	$^2P_{3/2}$		
2		$^2D_{3/2}$	$^2D_{5/2}$	
3			$^2F_{5/2}$	$^2F_{7/2}$

where the value of j is written as a subscript, and the superscript 2 indicates that the state is double, or a doublet. The new selection rules are $\Delta j = \pm 1, 0$, with the exception that the transition $j = 0 \rightarrow j = 0$ is ruled out. These considerations explain why the alkalis show a doublet spectrum. The series notation may now be written

$$n\,^2S_{1/2} - m\,^2P_{1/2}, \qquad n\,^2S_{1/2} - m\,^2P_{3/2},$$

Principal Series Doublets, etc.

The alkaline earths, in the second column of the periodic table, are divalent, and hence have two valence electrons situated at some distance from the atom core. To each electron is ascribed orbital angular momenta l_1 and l_2 and spins s_1 and s_2. In most cases one combines the two spins vectorially to give the resultant spin of the atom. Thus

$$\mathbf{s} = \mathbf{s}_1 + \mathbf{s}_2,$$

or $\qquad\qquad\qquad s = 0$ or 1.

The orbital angular momenta are also combined into a resultant orbital momentum

$$\mathbf{l} = \mathbf{l}_1 + \mathbf{l}_2,$$

where l takes all integral values from $|\,l_2 - l_1\,|$ to $|\,l_2 + l_1\,|$. The total angular momentum j is then found by combining l and s according to the scheme

$$\mathbf{j} = \mathbf{l} + \mathbf{s}.$$

For $s = 0$, it is clear that there is only one value of j for a given l, so that in this case the states are single, and single lines result from combinations of these states.

On the other hand, when $s = 1$, there are three values of j for every l (except $l = 0$), thus $j = l - 1, l, l + 1$, and triplet states result. By combination of these various triplet states with each other, a multiplicity of lines results, called multiplets. Some of the strongest lines in the alkaline earth spectra are due to a combination of triplet S states $(j = 1)$ and other triplet states resulting in the formation of triplet lines. The above theory explains the well-known fact that the alkaline earths exhibit spectra in which the strongest lines are singlets and triplets. The various states are designated by writing 1 or 3 as superscripts to the left of the letter designating l, to denote singlet and triplet states, thus

$$^1S_0, \quad ^3S_1, \quad ^3P_0, \quad ^3P_1, \quad ^3P_2, \text{ etc.},$$

and the lines are designated by

$$\left. \begin{array}{l} n\,^3P_0 \\ n\,^3P_1 \\ n\,^3P_2 \end{array} \right\} - m\,^3S_1, \text{ etc.}$$

Atoms containing more than two valence electrons may be built up in a manner similar to the one given above. In these cases various types of states, such as quartets, quintets, etc., may arise, depending on the number of external electrons in the atom. The multiplicity of the state is designated as above by putting a superscript before the symbol denoting the value of l.

The theory further predicts that the relative separation of the states j', j'', j''' of a given multiplet should be

$$\Delta\nu_{j'j''}:\Delta\nu_{j''j'''} = \{j'\,(j'+1)-j''(j''+1)\}:\{(j''+1)-j'''(j'''+1)\}.$$

This rule is known as the interval rule and is of great value in assigning the value of j to the various multiplet states. Another principle which is of importance in analysing spectra relates to the relative intensities of lines within a given multiplet. A simple statement, known as the sum rule of Ornstein, Burgers and Dorgelo, is as follows: *The sum of the intensities of all lines coming from a given upper level (j) of a multiplet state is proportional to $2j + 1$, the quantum weight of this state.*

2b. ENERGY LEVEL DIAGRAMS. It has been found convenient to make a diagram connecting the energies of various states with the quantum numbers of these states. Usually the energy of various states (see Fig. 1) is plotted as ordinate against the designation of the terms as abscissa. Thus the ^{1}S levels are usually plotted under each other, all ^{1}P levels under each other, etc., and the triplet levels are usually separated from the singlet levels. All observed spectral lines are represented by lines connecting the two energy states involved in the formation of the line. The energy levels are given on the right of the diagram in wave numbers (cm.$^{-1}$) with the ionization potential, or series limit, taken as the zero energy state. The lowest, or normal, state of the atom would therefore have the highest term value. In 1913 Franck and Hertz[*] showed that electrons, which had attained an energy $\frac{1}{2}mv^2$ by being accelerated through a potential difference V, lost no energy on collision with mercury atoms if their energy was less than 4·9 volts. If, however, they had been accelerated through 4·9 volts, a large fraction of them lost all their energy and the mercury vapour was found to emit the line[†] 2537. This experiment gave a definite proof of the existence of stationary states. It also created the custom

---

[*] J. Franck and G. Hertz, *Verh. d. D. Phys. Ges.* **15**, 34, 373, 613, 929 (1913); **16**, 12, 427, 512 (1914); **18**, 213 (1916).

[†] Whenever a spectral line is referred to by a number, this number will stand for the wave-length in Ångström units.

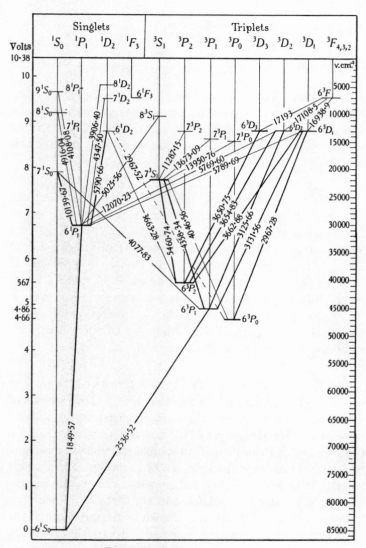

Fig. 1. Energy levels of mercury.

of expressing energy levels in volts. From the conservation of energy we have

$$eV = 1/2\ mv^2 = h\nu,$$

$$V = \frac{hc}{e\lambda} \text{ or } V = \frac{12336}{\lambda\ (\text{Ångströms})}.$$

In accordance with this scheme the energy in volts of the various states is also given on the energy level diagram, the zero point of measuring energy being taken as the lowest or normal state of the atom. These energies in volts are given on the left of the diagram.

2c. METASTABLE STATES. It will be noticed that there are certain states given on the diagram which are not joined to any other states by lines representing spectral lines, the reason being that the selection rules mentioned above do not allow electron jumps between the two states in question. If an atom is in such a state that it cannot jump to a lower energy state and emit radiation it is said to be in a *metastable* state. The atom must therefore stay in this state until it can give up its energy to another body by collision. The $6\,^3P_0$ and $6\,^3P_2$ states in the diagram are metastable states.

2d. NOTION OF MEAN LIFE OF AN EXCITED STATE. The existence of stationary states and the idea that the emission of a spectral line is due to the atom jumping from one stationary state to another at once raises the question as to how long, on the average, an atom stays in an excited state before returning to a lower state with the emission of radiation. The average length of time an atom stays in an excited state before returning to a lower state (if isolated and not subject to disturbing influences such as collisions) is known as the mean life of the atom in that state. As will be shown in Chap. III, the atom may be considered to have a certain probability of leaving an upper state n and jumping to various lower states m, emitting radiation. If the probability of a transition from a state n to m is A_{nm}, then the mean life is defined as the reciprocal of the sum

of the transition probabilities from the state n to all lower
states m. Thus

$$\tau_n = \frac{1}{\underset{m}{\Sigma} A_{nm}}.$$

The quantity A_{nm} is related to the intensity of the line of frequency ν_{nm}, and in some cases may be calculated theoretically.

These probability considerations now allow us to make a
slightly different interpretation of the selection rules. Instead
of stating that a transition from one state to another in which
$\Delta l = 0$, for example, is forbidden, it is more correct to say that
the probability of a transition between the two states is small.
If, now, an atom is in a so-called metastable state, it means that
any quantum jump it may make violates a selection principle.
The chance of leaving that state is therefore small and the
mean life long.

3. REMARKS ON FLUORESCENCE

It has long been known that certain solids and liquids, when
excited by monochromatic light of frequency ν, will themselves
emit light, usually a continuous spectrum of frequencies different from ν. Such a process is termed fluorescence. Early
work on gases showed that they exhibited fairly complicated
band fluorescence when excited by various types of radiation.
The fluorescence of sodium vapour was studied in the last
decade of the nineteenth century. When excited by sunlight
or light from a flame containing sodium chloride solution,
which emitted the sodium D lines(5890, 5896; $3\,{}^2S_{1/2}$-$3\,{}^2P_{3/2,\,1/2}$),
sodium vapour exhibited a system of bands which we now
know to be due to the Na_2 molecule. It was of course known
at the time that sodium vapour strongly absorbs the two
D lines, and furthermore classical theory predicted that
sodium vapour, illuminated by the D lines, should also
emit D lines. Several investigators, among them Wiedemann
and Schmidt[31], Wood[33, 34] and Puccianti[21], tried to find
this effect but were unsuccessful for various reasons.

Wood[33, 35], however, at a later date, succeeded in exciting
the D line fluorescence in sodium vapour by the action of the

D lines themselves. He vaporized some sodium in an evacuated test-tube and illuminated this with light from a gas flame containing NaCl solution, and observed a yellow fluorescence emerging from the tube from the point at which the exciting light entered. The cone of fluorescent light extended some distance back from the wall of the tube, but as the temperature of the test-tube was raised and the vapour pressure of sodium increased, the length of the fluorescent cone decreased, so that at high vapour pressures the fluorescent light was confined to the inner surface of the wall through which the exciting light entered. Spectroscopic investigation of the fluorescence showed that it contained only the two D lines. He termed this fluorescence *resonance radiation*, since it was predicted by the classical theory of a light wave vibrating with the same frequency as the dipole oscillations of the medium. It is clear from this experiment that the failure of the earlier attempts to find resonance radiation was due to the fact that the vapour pressure of sodium in the tube was too high.

The meaning of the experiment can be made clear on the basis of the quantum theory with the help of the energy level diagram of sodium (Fig. 2). The normal state of the sodium atom is the $3\,{}^2S_{1/2}$ state. When a continuous spectrum of light of wave-lengths from 2000 to 6000 is sent through a quartz cell containing sodium vapour, it is found that only the lines of the principal series are absorbed. This is to be expected, since at ordinary temperatures all sodium atoms in the tube are in the normal state and the principal series is the only series ending on the ground state. In Wood's experiment, atoms in the normal state absorbed the D lines and were thereby raised to the $3\,{}^2P_{1/2}$ and $3\,{}^2P_{3/2}$ states. (Other lines of the principal series did not pass through the glass walls of the tube.) Excited atoms in the $3\,{}^2P$ states then reverted to the $3\,{}^2S_{1/2}$ state, emitting the D lines as fluorescence.

As a definition one may say that if atoms in the normal state absorb light of a certain frequency, and subsequently re-emit light of the same frequency, the emitted light is termed *resonance radiation*. In terms of the energy level diagram it will be seen that resonance radiation will occur when an atom

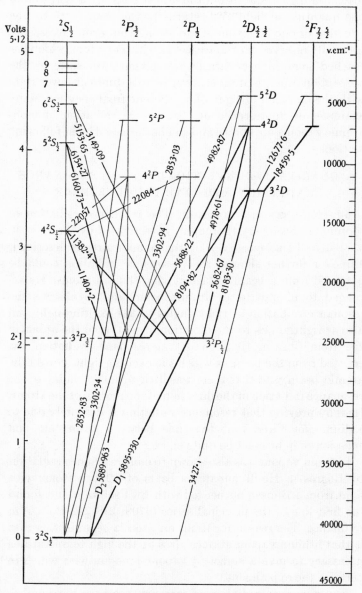

Fig. 2. Energy levels of sodium.

reaches a higher state from the lowest state by absorption of one quantum of light and returns to the same state by the emission of one quantum of radiation. The term fluorescence is usually reserved for those cases in which an atom, which has reached some higher state from a given lower state by the absorption of a quantum $h\nu$, returns to a different lower state with the emission of light of a different frequency ν_2. Many examples of fluorescence and resonance radiation of monatomic vapours are known and will be discussed in the following sections.

4. QUALITATIVE INVESTIGATIONS OF RESONANCE RADIATION AND LINE FLUORESCENCE

4a. RESONANCE RADIATION. The resonance radiation of sodium has been very thoroughly investigated by Wood[33, 34, 35, 39, 41] and Dunoyer[8, 9, 10] in the manner already described. Dunoyer further showed that resonance radiation could be obtained from a beam of fast-moving sodium atoms. He arranged to illuminate an atomic beam and to observe the resonance radiation at right angles to the exciting light and also at right angles to the direction of motion of the atoms in the atomic beam. He showed that resonance radiation was emitted from the point at which the exciting light crossed the atomic beam, and that there was little or no spreading of the resonance radiation in the direction of the motion of the atoms, thereby proving that resonance radiation is definitely due to sodium *atoms* and that the time between absorption and emission of light must be quite short.

Lithium vapour was also shown to emit resonance radiation by Bogros[3]. He illuminated a beam of lithium atoms with light from a Bunsen burner fed with LiCl solution and found the first line of the principal series (6708; $2\,{}^2S_{1/2}$–$2\,{}^2P_{3/2,\ 1/2}$) in resonance. The reason for using an atomic beam in this case is that lithium vapour attacks glass at the high temperatures necessary to give a sufficient vapour pressure with which to perform the experiment.

The resonance radiation of mercury has been very extensively studied by a great many investigators. Wood[37] first showed in 1912 that mercury vapour, at a vapour pressure corresponding to that at room temperature, when illuminated by light from a quartz mercury arc emitting an unreversed line (see Chap. I, §5) of wave-length 2537 ($6\,^1S_0$–$6\,^3P_1$), emitted this line as resonance radiation. Numerous experiments have been performed on the resonance line of mercury and they will be described in detail in the later sections of this book. The energy level diagram for mercury (Fig. 1) shows that the singlet line 1849 ($6\,^1S_0$–$6\,^1P_1$) should also be a resonance line, since it ends on the ground state. It was very difficult to obtain this line in resonance, as it is absorbed to a great extent by the oxygen of the atmosphere and also to some extent by the quartz apparatus needed to perform the experiment. Rump[23] was, however, finally successful in obtaining this line in resonance. His apparatus consisted of the usual quartz mercury arc and quartz resonance tube containing mercury vapour. The entire light path from arc to resonance tube to spectrograph was enclosed in tubes. Through these tubes, as well as the spectrograph, he circulated CO_2 gas, which does not absorb the line 1849. By this method he was able to show that the line actually appeared as resonance radiation.

The spectrum of cadmium and zinc is similar to that of mercury. Each element shows two resonance lines, a singlet line Cd 2288 ($5\,^1S_0$–$5\,^1P_1$); Zn 2139 ($4\,^1S_0$–$4\,^1P_1$) and an intercombination line Cd 3261 ($5\,^1S_0$–$5\,^3P_1$); Zn 3076 ($4\,^1S_0$–$4\,^3P_1$). Terenin[29] was able to obtain both lines 3261 and 2288 in resonance, when the exciting light source was a vacuum arc in cadmium. He was able to obtain good intensities of both lines at a vapour pressure of cadmium corresponding to 150° C. He further showed, by using filters to cut out the 2288 line from the source, that only 3261 appeared in resonance. He was also able to show that if the tube was excited by light containing 2288 and not 3261, only the former line appeared. By measuring the intensity of the 3261 line in resonance as a function of the vapour pressure of cadmium in the tube, he found that the intensity at first increased as the vapour pressure increased, reached a maximum, and then began to decrease with increasing vapour pressure. The maximum intensity appeared at a vapour pressure of 4×10^{-4} mm. The decrease in intensity of the resonance radiation obtained by increasing the vapour pressure beyond the point of maximum intensity is due to absorption of the 3261 line by cadmium atoms between the centre of the exciting beam and the window of the resonance tube through which the radiation is observed.

A similar experiment was performed with zinc vapour [20, 27, 32], showing that the two lines 2139 and 3076 appeared as resonance lines. Ponomarev and Terenin [20] showed that these two lines could be obtained when the vapour pressure of zinc was 5×10^{-4} mm., corresponding to a temperature of 280° C.

4b. RESONANCE RADIATION AND LINE FLUORESCENCE. The energy levels of a large number of atoms are such that they exhibit both resonance radiation and line fluorescence. In these cases the energy level diagram of the atom shows several low-lying states, all but one of which are metastable. Such an atom may absorb a given line from the source, thereby arriving at some higher level, in accordance with the selection rules, from which it may return either to the lowest state emitting resonance radiation or to one of the low-lying metastable states emitting fluorescent lines of longer wave-length than that of the line absorbed. The fluorescence of thallium vapour, first investigated by Terenin [29], is an example.

The energy level diagram of thallium (Fig. 3, showing only the lowest states) indicates that thallium has two absorption lines 3776 ($6\,^2P_{1/2}-7\,^2S_{1/2}$) and 2768 ($6\,^2P_{1/2}-6\,^2D_{3/2}$). If the atom absorbs 3776, it reaches the $7\,^2S_{1/2}$ state, from which it may revert to the normal state emitting the resonance line or may return to the metastable $6\,^2P_{3/2}$ state with the emission of the fluorescent line 5350. A similar situation arises if the atom has reached the $6\,^2D_{3/2}$ state by absorption of 2768, the two lines 2768 ($6\,^2P_{1/2}-6\,^2D_{3/2}$) and 3530 ($6\,^2P_{3/2}-6\,^2D_{3/2}$) being emitted. Terenin showed that, if the vapour, contained in a quartz tube at a vapour pressure corresponding to 450° C. to 500° C., be illuminated by light from a quartz thallium arc, the four lines mentioned above are re-emitted by the vapour. The interposition of a filter, transparent to the green but opaque to 3776 and lower wave-lengths, between the source and the resonance tube resulted in the obliteration of all fluorescence. Both lines, however, appeared when a filter transparent to 3776 but not to 5350 was used, thus showing the correctness of the assumed process.

Terenin also investigated the fluorescence of lead, bismuth, arsenic and antimony, the results of which are set forth in Table III. The case of antimony is an interesting one, since the metal readily forms molecules Sb_2 in the gaseous state. For

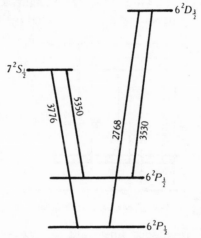

Fig. 3. Low-lying energy levels of thallium.

this reason the absorption spectrum was not well known, since in it the lines were confused with the molecular spectrum. Terenin investigated this case by the method of fluorescence, superheating the vapour to suppress molecule formation. (Furnace temperature 1100° C.; temperature controlling vapour pressure of antimony 200° C.–350° C.) A number of lines were seen in fluorescence. In order to distinguish between resonance lines and fluorescent lines and to tell which transitions were responsible for each, he developed the very ingenious method of "crossed spectra". He excited the fluorescence through a dispersive prism in such a way that the mono-chromatic images of the slit reached the resonance tube spread out in a vertical plane. The images of the fluorescent light from the tube were projected on to the vertical slit of a spectrograph, which was set up in the usual way with the direction of dispersion horizontal. A photograph of the fluorescence taken in this way showed a square array of "spectral points", spectral images of the different exciting beams entering the resonance tube. The spectrum of the fluorescence contained the lines 2878, 2770, 2598, 2528, 2671, 2311. On the photograph the lines 2878 and 2311 appeared to be excited by 2311; 2671 and 2770 by 2176; and 2598 by 2068. The other lines appearing on the plate were probably due to scattered light. The lines 2176 and 2068 did not appear on the plate, probably due to the fact

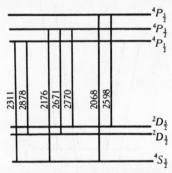

Fig. 4. Low-lying energy levels of antimony.

that they are resonance lines and are weakened by self-absorption. The energy states involved in the process have since been found to be those given in Fig. 4. It is a significant tribute to the power of the fluorescence method employed by Terenin that he was able to give a correct picture of the position of the lower energy levels of antimony at a time when the spectrum was little known and that later work has confirmed his results.

There remain to be discussed two cases in which higher series member lines of alkali atoms have been excited and emit resonance radiation together with certain fluorescent lines. The two elements are sodium, studied by Strutt[28], Christensen and Rollefson[7], and caesium, studied by Boeckner[2].

In examining the second members of the principal series of

sodium, 3302·34, 3302·94 ($3\,^2S_{1/2}$–$4\,^2P_{1/2,\,3/2}$), Strutt, and later Christensen and Rollefson, used a vacuum arc in sodium as a source. The arc was fitted with a quartz window and the resonance vessel was made of quartz, so that sodium vapour could absorb both the first and second members of the series. Usually a monochromator was used so that the vapour could be excited by the 3303 doublet alone. The fluorescent light was found to contain both the ultra-violet doublet and the D lines. The emission of the ultra-violet doublet is clearly a case of pure resonance radiation, whereas the emission of the D lines may be accounted for by a process of the following type. Atoms excited by the absorption of 3303 to the $4\,^2P_{1/2,\,3/2}$ states instead of returning to the normal state by the emission of 3303 (resonance radiation) may take the path

$$\text{(1)} \quad 4\,^2P \to 4\,^2S \to 3\,^2P \to 3\,^2S$$
$$\text{or} \qquad \text{(2)} \quad 4\,^2P \to 3\,^2D \to 3\,^2P \to 3\,^2S.$$

In either of these cascade-like emission processes the last step results in the emission of the D lines. The separate steps of either process (1) or (2) involve the emission of certain infrared lines. The probability of emission by route (1) relative to that by the direct route was found by Weiss[30 a] to be 25 to 1, but by Christensen and Rollefson[7] to be about equal. The reason for this discrepancy is at present unknown.

A similar case was studied by Boeckner[2] in caesium. Through the discovery that the strong helium line 3888 coincides with one of the doublet components of the third member of the principal series in caesium, 3888·6 ($6\,^2S_{1/2}$–$8\,^2P_{1/2}$), Boeckner was able to excite the caesium line by using a helium discharge tube as source. The advantage of this source is that it gives a strong exciting line exhibiting no self-reversal. The line 3888·6 ($6\,^2S_{1/2}$–$8\,^2P_{1/2}$) appeared in resonance, but the other member of the doublet, 3876·4 ($6\,^2S_{1/2}$–$8\,^2P_{3/2}$), did not, since it is so widely separated from the longer wave-length component as not to be excited by the helium line. Furthermore, the cascade process described above was also found to occur in caesium, since the first member of the series 8521·1 ($6\,^2S_{1/2}$–$6\,^2P_{3/2}$) was also found. Here, in contradistinction to the case of sodium, some of the intervening steps of the process were definitely shown. Since the lines 7609 ($6\,^2P_{1/2}$–$8\,^2S_{1/2}$) and 7944 ($6\,^2P_{3/2}$–$8\,^2S_{1/2}$) appear on the plate the most probable process appears to be $8\,^2P \rightarrow 8\,^2S \rightarrow 6\,^2P \rightarrow 6\,^2S$. The intensity of the resonance line 3888·6 was about the same as that of the fluorescence line 8521, indicating an equal probability for the direct and the cascade process.

5. SOURCES FOR EXCITING RESONANCE RADIATION

Before describing any further experiments on resonance radiation it may be well at this point to discuss the theory and construction of sources for exciting resonance radiation. To understand the properties of such sources the following considerations are necessary. The conception of strictly monochromatic light is a useful and necessary part of theoretical physics, but represents an abstraction that is impossible of experimental realization. The light emitted by a group of motionless atoms, far apart, constitutes a wave train that is continually suffering interruptions, and has therefore a finite spectral width. The added effect of the motions of the atoms and their mutual interactions is to produce a spectral line, that is, a distribution of light intensity over a range of wavelengths which has a maximum at some particular wave-length (designated as the wave-length of the spectral line) and grades off to zero on both sides of the maximum. The portion of the line in the immediate neighbourhood of the maximum we shall designate roughly as the centre of the line, and the further portions as the edges.

If the light from a group of excited atoms could issue from a tube without further ado, all such sources of light would be equally suitable for the excitation of resonance radiation. This, however, is unfortunately not the case. In the usual source of so-called "monochromatic" light, such as a flame or an arc, there is a central hot portion where most of the electrical or thermal excitation is taking place, and an outer, cooler, less excited portion which is capable of absorbing the light emitted by the central part. Even this would not be serious were it not for the fact that this outer portion is capable of absorbing the centre of the line to a much greater extent than the edges. This phenomenon will be considered in great detail later, but it is sufficient at this point to emphasize that the line emitted from a source under these conditions consists of a distribution of intensity in which there are two maxima on either side of the centre of the line, and a minimum at the centre. Such a line is said to be "self-reversed", and the phenomenon is called "self-reversal". If the line has a hyperfine structure, then each component exhibits self-reversal. For example, an analysis by an instrument of high resolving power of the 2537 line from an ordinary mercury arc indicates ten separate maxima, which can be shown to be five self-reversed hyperfine structure components.

Since the outer portion of an arc consists mainly of normal atoms, only spectral lines terminating at the normal state will exhibit marked self-reversal. Consequently self-reversal is a convenient aid to the spectroscopist in recognizing resonance lines. When it is desired, however, to excite resonance radiation in a group of unexcited atoms, it is essential that the *exciting light be unreversed*, inasmuch as precisely that missing portion of a self-reversed line is the portion that is effective. We have therefore two general conditions that must be fulfilled by a lamp which is to be used for exciting resonance radiation, namely: (1) the resonance lines must be intense, and (2) the resonance lines must show no self-reversal.

5a. ARCS WITHOUT FOREIGN GAS. The first arc ever designed for exciting resonance radiation, and one that is still used to a great extent to-day, is the water-cooled, magnetically deflected quartz mercury arc developed by Wood. In its present form it consists of an ordinary quartz mercury arc of the vertical type (manufactured by Heraeus, Hanovia, General Electric Vapour Lamp Company, etc.) with the cathode end immersed in water. A weak magnetic field is used to press the arc stream (and therefore the emitting layer) against the front wall, and a large inductance and a resistance are put in series with the lamp, the former to keep fluctuations as small as possible and the latter to keep the current as low as possible. Although this lamp is very effective in exciting resonance radiation (2537) in mercury vapour, it has the disadvantage of not being steady, particularly when running at low current. A similar arc containing an alloy of cadmium and tin has been used with some success by Bates[1] to excite cadmium resonance radiation (3261).

Mitchell[16] used an arc with a hot cathode in zinc vapour for producing the resonance lines which were capable of exciting zinc resonance radiation. An outside oven was used to vaporize the zinc, and a voltage of 110 volts was established between the oxide-coated cathode and the plate. With a plate current of about 5 amperes, the zinc spectrum was very intense. With such a lamp care must be taken that the plate does not sputter the inside walls of the tube and diminish their transparency, and also, to avoid self-reversal, the layer of unexcited vapour lying between the emitting layer and the exit window must be made as small as possible.

5b. ARCS WITH STATIONARY FOREIGN GAS. It is an important result of the researches of many investigators that the ions, electrons and metastable atoms present in an inert gas arc discharge are very effective in exciting the arc lines of a metallic vapour mixed with the inert gas. The phenomenon is most surprising. The inert gas carries the discharge, which remains quite constant, but the spectral lines emitted belong almost entirely to the admixed vapour. For example, when mercury vapour is present in a helium discharge, the helium lines are very faint, and the mercury lines very strong. This fact has been used in five different ways in the construction of sources for exciting resonance radiation.

The first tube employing this principle was described by Ellett[11] and used by him to excite sodium resonance radiation. In the hands of others it has been used to excite resonance radiation in the vapours of Zn, Cd, Cs, K, Tl and Hg. In the Ellett tube a high voltage discharge (about 3000 volts) between two hollow cylindrical electrodes is maintained across hydrogen which streams slowly from one end of the tube to the other. Sodium vapour, issuing from some solid sodium contained in a side tube and heated by an outside oven, enters the hydrogen stream at about the middle of the tube. At this point a sodium glow appears, which is effective in exciting sodium resonance radiation.

In the Schüler[26] tube, a hollow cylinder of nickel is used as a cathode and a small amount of the substance whose arc spectrum is to be investigated is placed inside. The tube is filled with an inert gas, and a high voltage discharge is produced. The cathode is heated by positive ion bombardment to a sufficient extent to vaporize the substance within, and the vapour thus formed is excited by collisions with the inert gas atoms and ions. To prevent the cathode from becoming too hot it is cooled by water or in some cases by liquid air. This type of tube was used by Mitchell[16] to produce the zinc resonance lines which were found to be quite effective in exciting zinc resonance radiation. Inasmuch, however, as there are simpler types of tubes that are just as good for exciting resonance radiation, the Schüler tube is not used very much for this purpose, being reserved mainly for hyperfine structure investigations where, because of the liquid air cooling, it gives extraordinarily sharp lines.

Kunze's[15] tube differs from the Schüler tube only in that the substance whose vapour is to be excited is contained in a side tube where its temperature (and hence the vapour pressure) can be accurately controlled. The tube is filled with a few millimetres of an inert gas and a discharge of from 220 to 240 volts is established across two hollow cylindrical iron electrodes by using a high frequency spark to start the discharge. Between the electrodes the tube is constricted into a short capillary, from which most of the light is emitted. When very great intensity is not desired this lamp is to be recommended, but if the current is made large, the current density in the capillary may become large enough to broaden the resonance line and thereby reduce the intensity at its centre.

A very intense source of practically unreversed light has been devised by Pirani[19] and has been placed on the market. A schematic diagram is shown in Fig. 5. An inner tube containing solid sodium and a few millimetres of an inert gas is

fitted with two oxide-coated filaments. This is surrounded by
a second tube and the intervening space is evacuated. To start
the lamp a current of a few amperes is sent through both
filaments for a moment until they are red hot, and then this
current is shut off at the same time that an alternating voltage
of about 110 volts is established across the two filaments. An

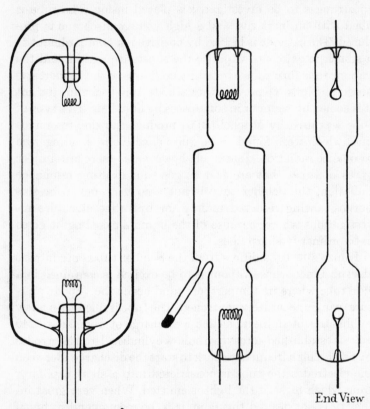

End View

Fig. 5. The Pirani lamp.   Fig. 6. The Houterman's lamp.

arc strikes and keeps the filaments hot without the necessity of
sending current through them. As the inner tube warms up,
the sodium vaporizes and the whole inner tube glows with
sodium light. Because of the surrounding vacuum, the inner
tube is at practically a uniform temperature, and very little
self-reversal is produced. The Pirani lamps have been made

for cadmium and magnesium as well as sodium, and will presumably work with other materials as well.

.Houtermans [14] has recently described a modification of the Pirani lamp in which the necessity of the outer tube has been obviated. The central portion of the tube has been constricted to the form of a flat slab about two millimetres thick (see Fig. 6). Because of its thinness this slab of glowing vapour is at practically a uniform temperature, and measurements of Houtermans indicate that practically no self-reversal takes place. The substance to be vaporized is placed in a side tube and kept at a known temperature by an outside oven. With mercury, the side tube was placed in water. The lamp is filled with from three to four millimetres of argon and a potential of about 180 to 200 volts A.C. is used. (The discharge is first started with a "high frequency spark".) The current should not exceed 3·6 amperes. Various tests indicate that the breadth of the lines emitted by this lamp is determined entirely by the Doppler effect, and consequently, once the temperature and the absorption coefficient of the emitting vapour are known, the actual line form can be calculated (see Chap. III, § 3a). With mercury in the side tube, both resonance lines 2537 and 1849 were obtained, whereas with magnesium, the inter-combination resonance line 4571 was missing, perhaps because of impurities that were present.

5c. ARCS WITH CIRCULATING FOREIGN GAS. In the lamps of Kunze and of Houtermans, the material whose vapour is to be mixed with the foreign gas is contained in a side tube, the temperature of which is controlled from the outside by suitable cooling or warming devices. With mercury, these lamps are particularly effective because a satisfactory vapour pressure is obtained when the side tube is immersed in water at room temperature. Under these conditions no mercury condenses on the walls of the tube proper because these walls are maintained at a higher temperature by the arc discharge. In using such lamps with materials which have to be heated to a high temperature to give sufficient vapour, it is necessary to run the arc at a high current to prevent condensation on the exit

window. It is not expedient, however, to use a high current
because of the attendant broadening of the lines. It is therefore worth while to construct an arc to be operated at low
current, allowing the portion of the tube in the neighbourhood
of the exit window to remain cool with no condensation, however, taking place on these cool surfaces. This is achieved by
circulating an inert gas (across which an arc discharge is
established) in a direction *from* the exit window *toward* the
stream of hot vapour which issues from the heated solid or
liquid. The first arc of this type was invented by Cario and
Lochte-Holtgreven [5]. In its present form, developed by
Ladenburg and Zehden [42], it can be operated on 220 volts D.C.

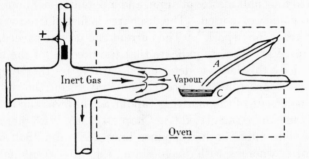

Fig. 7. The Cario-Lochte-Holtgreven lamp.

and with a current of from 80 to 100 milliamperes. A schematic
diagram of the improved Cario-Lochte-Holtgreven lamp is
given in Fig. 7. The metal, whose resonance lines are desired,
is distilled through the tube A into the cathode C, which is a
small iron boat. From three to five millimetres of an inert gas
are introduced and circulated in the direction of the arrows by
a circulating pump. The iron cathode is first outgassed by a
heavy discharge, and then fresh gas is introduced. This may
have to be done several times. Once the system is outgassed the
tube should run on a voltage around 220 volts, provided the
discharge is started with a high frequency spark. The oven,
indicated by dotted lines, can be made as hot as the glass tube
will stand.

The features of this lamp are as follows: (1) No matter how
hot the oven is, the exit window remains at room temperature

and can therefore be sealed to the tube with ordinary cement. (2) The emitting layer of vapour is in the hottest part of the oven and is practically at a uniform temperature. (3) There is no layer of unexcited vapour between the emitting layer and the exit window. (4) The current can be varied by an external rheostat and the vapour pressure can be varied by varying

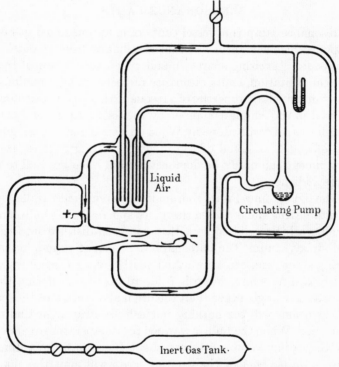

Liquid Air

Circulating Pump

+

Inert Gas Tank·

Fig. 8. Connection for Cario-Lochte-Holtgreven lamp.

the oven temperature, both adjustments being independent. (5) Once the tube is operating properly no adjustments are necessary and no deterioration occurs. (With sodium in the tube, Zehden operated this lamp for more than two years without the slightest deterioration.)

In Fig. 8 a diagram of the connection to the circulating pump is given.

The modification of the Cario-Lochte-Holtgreven lamp

described by Christensen and Rollefson[6] has the objection that the emitting layer of vapour is half in the oven and half out, so that a temperature gradient exists which must produce some self-reversal. The emitting layer also is more than 10 cm. long, whereas in the Ladenburg-Zehden modification it is only about 3 cm. long.

6. RESONANCE LAMPS

A resonance lamp is a vessel containing an unexcited gas or vapour which can absorb a beam of radiation from an outside source (an "exciting source"), and which, as a result of this optical excitation, emits resonance radiation in all directions. The fundamental properties of a resonance lamp can be demonstrated in the simplest manner by the following experiment, which was originally done by Wood on a small scale. Owing to the efficiency of some of the modern exciting sources, this experiment can easily be demonstrated in a lecture hall to a large audience.

The light from a Pirani sodium lamp is rendered parallel by a suitable lens and is passed through a spherical glass bulb containing a little solid sodium that has been distilled into the bulb in a vacuum. The bulb can be conveniently mounted on a ring stand, and can be warmed gently by a bunsen burner. When the resonance lamp is cold, an observer, viewing the bulb at any angle other than the original direction of the exciting beam, will see nothing but a little stray light due to reflection. When the bulb is warmed to a temperature of about $80°$ C. (sodium vapour pressure about 10^{-7} mm.), the sodium vapour *in the path* of the exciting beam will emit the characteristic yellow resonance radiation, and there will be no resonance radiation coming from any other part of the bulb. As the sodium vapour pressure is increased, the resonance radiation emitted from that part of the vapour which lies in the path of the exciting beam will increase until the vapour pressure rises to a value in the neighbourhood of 10^{-4} mm., at which moment the *whole* bulb begins to glow with resonance radiation. When the vapour pressure is made still higher, only the portion of the vapour lying near the window where the exciting beam enters is luminous, and if the vapour pressure is increased still further no atomic resonance radiation is emitted at all, only a band radiation associated with Na_2 molecules. As the bulb cools the phenomena take place in reverse order, until once again the luminous part of the vapour is confined to the path of the exciting beam.

These phenomena will be discussed in detail in later portions of the book. It is sufficient at this time merely to point out that the resonance radiation emitted by atoms in the direct path of the exciting beam is the result of a single atomic absorption and emission, and, if further absorptions and emissions by

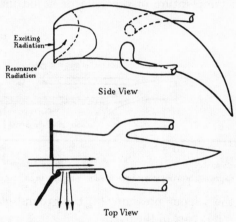

Fig. 9. Resonance lamp.

atoms not in this region (which cause the whole bulb to glow) are to be avoided, the vapour pressure must be kept low. From the standpoint of design, there are two main disadvantages of the bulb just described, namely: (1) stray light due to reflection is always present to some extent, and (2) there is always a layer of unexcited vapour lying between the path of the exciting beam and the window from which the resonance radiation emerges. Both of these defects are eliminated in the lamp depicted in Fig. 9, which embodies the best features of lamps developed by Wood, Kunze and Zehden. The light trap prevents internal reflections, and the slight projection of the entrance window ensures that the exciting beam will graze the exit window. When it is not necessary to place the resonance lamp in an oven (for example when working with mercury vapour), the entrance and exit windows may be cemented on the tube; otherwise the whole lamp must be blown of glass or of quartz.

When temperatures above room temperature are required the method of using such a resonance lamp is shown in Fig. 10, in which the following points are to be emphasized: (1) the whole resonance lamp must be placed at the hottest part of the oven, that is, at the centre, (2) the side tube containing the material to be vaporized must extend to a cool part of the oven, so that the temperature of the solid or liquid material will

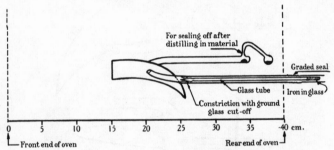

Fig. 10. Diagram showing use of resonance lamp at high temperatures.

determine the vapour pressure, (3) the ground glass cut-off must be in a hot part of the oven.

When the vapour pressure in the resonance lamp is so low that only atoms in the path of the exciting beam emit light, the frequency distribution of the emitted light is determined chiefly by the Doppler effect (see Chap. III, §3a), being slightly broader than the Doppler breadth. A resonance lamp capable of emitting a line with a breadth equal to the natural breadth of the line (which is usually much narrower than the Doppler breadth) has been constructed by Thomas[30]. Instead of using stationary mercury vapour whose atoms have a Maxwellian distribution of velocities, Thomas used an atomic beam. With the exciting beam and the atomic beam perpendicular, the resonance radiation taken off along the third perpendicular direction was uninfluenced by the Doppler effect. A similar device was constructed by Schein [24] and used as an absorption cell, and by Bogros [4] for measurements of lithium resonance radiation.

In determining the vapour pressure in a resonance lamp, the temperature of the solid or liquid in the side tube is determined by a good thermometer or thermocouple, and the vapour pressure p corresponding to this temperature T is calculated from the vapour pressure equation. If T' is the temperature of the centre of the oven, where the resonance lamp is situated, then the desired vapour pressure p' is given by

$$\frac{p'}{p} = \sqrt{\frac{T'}{T}},$$

and the number of atoms per c.c. in the resonance lamp, N', is given by

$$p' = N'kT'.$$

7. RESONANCE RADIATION AND SPECULAR REFLECTION IN MERCURY VAPOUR

Soon after his discovery of the resonance radiation of mercury, Wood [36, 37, 38, 40] and his students made a quantitative study of the intensity of the line 2537, appearing as resonance radiation, as a function of the vapour pressure of mercury in the resonance tube. They focused the image of the slit of a monochromator, set to pass 2537, on the window of a quartz bulb containing mercury vapour, and arranged to photograph the bulb with a camera fitted with a quartz lens. Since the resonance bulb contained a drop of liquid mercury and was then evacuated and sealed off, the vapour pressure of mercury could be controlled by placing the bulb in a furnace. At temperatures from 20° C. to about 150° C. a phenomenon was found similar to that observed with sodium and discussed in § 6. At low temperatures the resonance radiation filled the whole bulb, and as the vapour pressure was increased the luminous volume gradually contracted and at temperatures of about 150° C. the resonance radiation was emitted from a small layer adjacent to the entrance wall of the tube and was limited in size to about the dimensions of the portion of the window illuminated (the image of the monochromator slit). Above this temperature the image of the slit, formed at the mercury vapour surface, became very sharp and could only be seen in that direction corresponding to specular reflection. The diffusely scattered radiation reached a maximum at a temperature of about 100° C. (vapour pressure 0·3 mm.). At 150° C. (3 mm.) it decreased to one-half the maximum intensity, at 200° C. (18 mm.) to one-quarter, and at 250° C. (76 mm.) to about one-tenth. At 270° C. there was no trace of diffuse scattering, the entire resonance light being in the specularly reflected beam. A special type of resonance bulb was used in these experiments. It consisted of a thick-walled quartz vessel with a prismatic

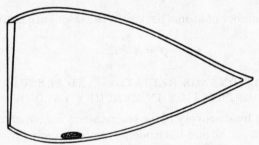

Fig. 11. Bulb for obtaining specular reflection.

window sealed to the front. This window served to separate the image of the slit reflected from the front (quartz) surface from that reflected by the mercury vapour.

The explanation offered by Wood was as follows. At low pressures a mercury atom may absorb light from the source, become excited, and finally re-emit the light as resonance radiation. In this case the light will be emitted uniformly in all directions. At high pressures (high temperatures) a similar process takes place, but at these pressures the atoms are very close together and it may be possible that the spherical waves emitted by these atoms have their phases so related that radiation occurs in only the direction of the specularly reflected beam.

The case of the specular reflection from mercury vapour has been found to be very similar to that of metallic reflection. Wood showed that, if the exciting beam were polarized, the specularly reflected beam would in general be elliptically polarized, as in the case of metallic reflection. Rump [22] further showed that the form of the 2537 line which was specularly reflected is independent of the temperature of the mercury vapour. To do this he illuminated the bulb containing the mercury vapour at high temperature with light from a resonance lamp containing mercury vapour at room temperature. It is well known that, if a resonance lamp R_{II}, containing mercury vapour at a pressure p and temperature T, is illuminated by light from another resonance lamp R_I containing mercury vapour at the same pressure but at a different temperature, the form of the line (Doppler breadth) is dependent only on the temperature in R_{II}. Rump showed this by measuring the form of the line (absorption measurement) emitted by mercury vapour in a resonance lamp R_{II}, which was so arranged that the vapour pressure of the mercury could be kept constant and the temperature of the vapour changed. When the bulb R_{II} was illuminated by light from the lamp R_I, it was found that the form of the line emitted by R_{II} was dependent on the temperature of the vapour in R_{II} if the vapour pressure was low. On the other hand at high pressures, where specular reflection is predominant, the form of the line emitted by R_{II} is the same as that of the exciting line (from R_I) and does not depend on the temperature of the vapour. Rump further showed that the form of a line specularly reflected from a metal surface is the same as that of the incident line. This brings out further the analogy between specular reflection from mercury vapour and metallic reflection. There is, however, one important difference, namely, that in mercury vapour specular reflection is highly selective, only wave-lengths in the immediate neighbourhood of 2537 being reflected, whereas this is not true of the usual metallic reflection.

That the processes of emission and absorption in dense vapour giving rise to specular reflection are not the same as those involved in the production of resonance radiation in rarefied gases was shown by Schnettler [25]. He investigated the quenching effect of hydrogen and carbon dioxide on the intensity of the specularly reflected beam. It is well known that the intensity of resonance radiation emitted by mercury vapour at room temperature can be considerably reduced by the admission of $0 \cdot 2$ mm. of H_2 or 2 mm. of CO_2. Schnettler showed that at high pressures ($370°$ C.) large amounts of H_2 and CO_2 decreased the intensity of the specularly reflected light by only a small fraction, the decrease per collision being about the same for each gas. This would imply that the mean life of the atom plays no rôle in the process of specular reflection.

8. HYPERFINE STRUCTURE OF LINE SPECTRA

So far, we have discussed those characteristics of line spectra which can be observed with spectroscopic apparatus of ordinary resolving power. It was discovered some years ago that, if certain lines of the spectrum of mercury or manganese were observed with apparatus of high resolving power such as a Lummer-Gehrcke plate or a Fabry-Perot étalon, these lines showed an extremely fine-grained structure, which has since been termed hyperfine structure. For example, if the 5461 line of mercury, coming from a well-cooled arc lamp, is observed with a Fabry-Perot étalon and a spectrograph, it exhibits a number of components. Originally the strongest component was assigned the wave-length corresponding to the line in question, and the wave-lengths of the weaker components, termed satellites, were determined with respect to this line. The order of magnitude of the separation of the various components ranged from a few tenths to a few hundredths of a wave number.

The elements which were first found to exhibit h.f.s.* all had many isotopes. This led to the assumption that the h.f.s. splitting was due to the change in mass of the nucleus. A short calculation showed, however, that the splitting to be expected on this assumption was of a much smaller order of magnitude than that observed. Furthermore, Goudsmit and Back† observed h.f.s. in the lines from the element bismuth, which consists of only one isotope.

---

* Hereafter "h.f.s." will stand for "hyperfine structure".
† S. Goudsmit and E. Back, *Z. f. Phys.* **43**, 321 (1927).

W. Pauli, Jr.[*] remarked that h.f.s. could be explained by assuming that the nucleus has a resultant spin $\dfrac{i}{2}\dfrac{h}{\pi}$ and a corresponding magnetic moment. On the basis of this hypothesis it is possible to account for the various h.f.s. levels with the aid of the vector model of the atom [†]. Consider an atom of an element with only one isotope (for the sake of simplicity) in a given energy state characterized by the vectors l, s and j. The nucleus may have a spin i which is an even or odd multiple of $\dfrac{1}{2}\dfrac{h}{2\pi}$ and which is the same for any nucleus of a given isotopic kind, and differs from one type of atom to the next. The total angular momentum may then be obtained by the vector sum

$$\mathbf{j} + \mathbf{i} = \mathbf{f}.$$

The new quantity f is called the hyperfine quantum number and may take all values in the range

$$|j + i| \geqslant f \geqslant |j - i| \qquad \qquad \ldots\ldots(3).$$

For an atomic state in which $j \geqslant i$ there are $2i + 1$ values of f; for $i \geqslant j$, on the other hand, there are $2j + 1$ values. If one considers the interaction energy of (a) the motion of the orbital electron (considered as a current) on the magnetic nucleus (considered as a little magnet), and (b) that of the spin of the electron on the spin of the nucleus (the interaction of two small magnets), one can find the relative separation between two states with different values of f. In transitions from one h.f.s. state to another the following selection rules exist:

$$\Delta f = \pm 1, \quad 0,$$

but $f = 0 \rightarrow f = 0$ is forbidden.

The above considerations give correct values for the separations of the various h.f.s. components if it be assumed that resultant spin i be due to the spin of a proton in the nucleus, and the magnetic moment connected with the said spin is $g(i)\dfrac{eh}{4\pi mc}$, where $g(i)$ is a very small number.

[*] W. Pauli, Jr., *Naturwiss.* **12**, 741 (1924).

[†] For a complete discussion of the phenomenon of h.f.s. see H. Kallmann and H. Schüler, *Ergeb. der Exakten Naturwiss.* **11**, 134 (1932).

When one studies elements with several isotopes one finds that isotopes of even atomic weight usually have a zero spin, with the exception of nitrogen which has a spin of 1. Isotopes

TABLE I

SOME NUCLEAR SPINS (KALLMANN AND SCHÜLER)

Z	Element	Isotopes	Spin
1	H	1	1/2
2	He	4	0
3	Li	6	0
		7	3/2
7	N	14	1
8	O	16	0
9	F	19	1/2
11	Na	23	3/2
15	P	31	1/2
17	Cl	35	5/2
25	Mn	55	5/2
29	Cu	63, 65	3/2
31	Ga	69, 71	3/2
33	As	75	3/2
35	Br	79, 81	3/2
37	Rb	85, 87	3/2
48	Cd	111, 113	1/2
		110, 112, 114, 116	0
49	In	115	5/2
51	Sb	121, 123	3/2 (?)
53	J	127	9/2
55	Cs	133	7/2 (?)
56	Ba	137	3/2 (?)
		136, 138	0
57	La	139	5/2
59	Pr	141	5/2
75	Re	187, 189	5/2
79	Au	197	3/2 (?)
80	Hg	199	1/2
		201	3/2
		198, 200, 202, 204	0
81	Tl	203, 205	1/2
82	Pb	207	1/2
		204, 206, 208	0
83	Bi	209	9/2

of odd atomic weight usually exhibit a spin which is an odd multiple of $\dfrac{1}{2}\dfrac{h}{2\pi}$, as is shown in Table I.

In the case of an element consisting of several isotopes of odd and even atomic weights one would expect that all lines due to isotopes of even atomic weight would coincide, since the net spin of these is zero. It is found, however, that this is not always the case, and that the lines coming from isotopes of even atomic weight may be displaced with respect to each other by amounts as large as the displacements due to nuclear spin. To explain this so-called "isotope shift", one may assume that the electric field in the neighbourhood of the nucleus of one isotope is different from that in the neighbourhood of another. In general, the h.f.s. pattern exhibited by a line from an element consisting of a number of isotopes may be quite complicated. It often happens that lines from isotopes of even atomic weight, having no nuclear moment, coincide, due to the isotope displacement, with lines from isotopes of odd atomic weight which exhibit a nuclear spin i.

In analysing a given h.f.s. pattern the following methods are used. In the first place the separations of the various h.f.s. components of various lines are measured as exactly as possible. The isotopic constitution and relative abundance of the various component isotopes must be known. These data, obtained with the help of a mass spectrograph, are usually at hand. For those elements consisting of a single isotopic species, the spin may be obtained with the help of interval rules and intensity formulae, similar to those used for multiplets showing ordinary multiplet structure. For elements consisting of several isotopes the intensity formulae and relative abundance of isotopes usually enable one to work out the pattern.

As an example of the structure of a line emitted by an atom consisting of several isotopes we shall discuss briefly the h.f.s. of the resonance line 2537 of mercury, which was first observed by Wood* and correctly explained by Schüler† and his collaborators. The h.f.s. pattern consists of five lines, as shown in Fig. 12. The lines coming from the isotopes of even atomic weight are marked X, together with the atomic weight of the isotope from which they come. The level diagram for the h.f.s. states of the two odd isotopes is given in Fig. 13. The isotope

---

* R. W. Wood, *Phil. Mag.* **50**, 761 (1925).

† H. Schüler and J. Keyston, *Z. f. Phys.* **72**, 423 (1931); H. Schüler and E. G. Jones, *ibid.* **74**, 631 (1932).

of atomic weight 199 exhibits a spin $i = 1/2$, while that of 201 shows $i = 3/2$. The lines on Fig. 12 are lettered to correspond to those on Fig. 13, and their relative intensities are given beside each line. The relative abundance of the various isotopes as

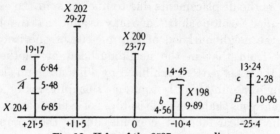

Fig. 12. H.f.s. of the 2537 mercury line.

$$\lambda = 2537\,(6\,^1S_0 - 6\,^3P_1)$$

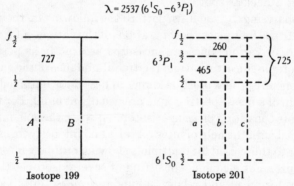

Fig. 13. Hyperfine level diagram for 2537.

given by Aston's mass spectrograph measurements is shown in Table II.

TABLE II

$Hg_{204} = 6.85\%$
$Hg_{202} = 29.27\%$ $Hg_{201} = 13.67\%$
$Hg_{200} = 23.77\%$ $Hg_{199} = 16.45\%$
$Hg_{198} = 9.89\%$
$Hg_{196} = 0.10\%$
Even Isotopes 69.88 %

It will be noted from Fig. 12 that the components $+11.5$ and 0.0 mÅ. have the relative intensity $29.27 : 23.77$, so that they must be due to the isotopes 202 and 200 respectively.

These two isotopes show no spin but are separated, due to the "isotopic displacement" effect. The line $+21\cdot5$ mÅ. consists of three components superimposed on each other, whereas the lines $-10\cdot4$ and $-25\cdot4$ mÅ. each consist of two components superimposed. It will be further observed that the sum of the components marked X is $69\cdot88$, the sum of $A+B$ is $16\cdot45$ (corresponding to Hg_{199}), and that of $a+b+c$ is $13\cdot67$ (corresponding to Hg_{201}). Furthermore, the relative intensities of $A:B$ and of $a:b:c$ are given by the usual intensity formulae.

9. INVESTIGATIONS ON THE HYPERFINE STRUCTURE OF RESONANCE RADIATION

Of the various elements exhibiting h.f.s., the resonance radiation of only one, mercury, has been studied with high resolving power apparatus. Ellett and MacNair[12] (see Chap. v) were the first to show that, if a resonance tube containing mercury vapour was illuminated by light from a well-cooled mercury arc, the resonance radiation emitted therefrom, when examined with a Lummer plate, showed all five h.f.s.components.

More recently, Mrozowski[17], in a long series of researches, has made a very thorough investigation of the h.f.s. of the resonance line of mercury. He observed that if light from a mercury arc was sent through a cell containing mercury vapour in a strong magnetic field, certain of the h.f.s. components of the resonance line could be filtered out, due to the Zeeman effect exhibited by the various components. By a proper choice of magnetic field he was able to let through four groups of lines: (α) the $-25\cdot4$ mÅ. component, (β) the $-10\cdot4$ mÅ. and $+21\cdot5$ mÅ. components, (γ) the $0\cdot0$ mÅ. and $+11\cdot5$ mÅ. components, and (δ) all five components. By using these various combinations of h.f.s. components as an exciting source to obtain resonance radiation from a tube containing mercury vapour, he found that the resonance radiation excited by a given group of h.f.s. components showed only those components contained in the exciting source. For example, the resonance radiation excited by the $-25\cdot4$ mÅ. component contained only that component, and that excited by the two components $-10\cdot4$ mÅ. and $+21\cdot5$ mÅ. contained these two components, etc. This proves that each h.f.s. line is itself a resonance line and that no line fluorescence occurs. This is exactly what would be expected from the energy level diagram of the previous section. Thus, excitation by the 0·0 and + 11·5 mÅ. components (case γ) leads to the excitation of the isotopes of mass 200 (absorption of the 0·0 component) and of mass 202 (absorption of the + 11·5 component). Since the two isotopes act as two independent gases, and since each has but one lower h.f.s. state, it is clear that each component must behave as a resonance line. Similar considerations apply to all the other components.

Another group of experiments performed by Mrozowski [18] has to do with the effect of added gases on the h.f.s. of the resonance radiation of mercury. As will be explained in Chap. II, collision between an excited atom and a foreign gas molecule may result in the transfer of the former from one excited level to another. If light containing the components 0·0 and + 11·5 mÅ. was allowed to excite mercury vapour containing a little helium, it was found that the resonance radiation contained only these two lines. On the other hand, if the excitation was by the − 25·4 mÅ. component, the three components − 25·4, − 10·4 and + 21·5 mÅ. were found in resonance. The explanation of these experiments follows at once from Fig. 13. The two lines 0·0 and + 11·5 mÅ. are due to the two isotopes 200 and 202 respectively, each exhibiting but one upper and one lower level. It follows, then, that a foreign gas can have no effect on the number of components obtained by this method of excitation. The component − 25·4 mÅ. consists of lines from the two isotopes 199, which have two upper levels $f = 3/2$ and $1/2$, and 201, having three upper levels $f = 1/2, 3/2, 5/2$. Excited mercury atoms of one isotopic kind may be transferred from one upper state f to other upper states. Thus collisions of excited atoms of isotopic mass 199 will result in the excitation of the $f = 1/2$ as well as the $f = 3/2$ states, giving rise to the lines A and B, contributing to the + 21·5 mÅ. and the − 25·4 mÅ. components respectively. Atoms of mass 201 may, by collision, arrive in the state $f = 5/2$ and $f = 3/2$ as well as $f = 1/2$, giving rise to the lines a, b, c, contributing to the components + 21·5,

TABLE III. RESONANCE AND FLUORESCENT LINES; METHODS OF PRODUCTION

Element	Observed resonance lines	Accompanying fluorescent lines	Temperature controlling vapour pressure	Exciting source	References
Na	$5889 \cdot 9\ (3\,^2S_{1/2}-3\,^2P_{3/2})$ $5895 \cdot 9\ (3\,^2S_{1/2}-3\,^2P_{1/2})$	—	80–150° C.	Sodium flame Vacuum arc Discharge tube	Wood [33, 35] and others
Na	$3302 \cdot 9\ (3\,^2S_{1/2}-4\,^2P_{3/2})$ $3302 \cdot 3\ (3\,^2S_{1/2}-4\,^2P_{1/2})$	$5890\ (3\,^2S_{1/2}-3\,^2P_{3/2})$ $5896\ (3\,^2S_{1/2}-3\,^2P_{1/2})$	80–150° C.	Vacuum arc Discharge tube	Strutt [28], Christensen and Rollefson [7]
Li	$6708\ \begin{cases}(2\,^2S_{1/2}-2\,^2P_{1/2})\\(2\,^2S_{1/2}-2\,^2P_{3/2})\end{cases}$	—	Molecular beam Boiler–500° C.	Flame (LiCl)	Bogros [3]
Cs	$3886\ (6\,^2S_{1/2}-8\,^2P_{1/2})$ $[3876\ (6\,^2S_{1/2}-8\,^2P_{3/2})]$	$7609\ (6\,^2P_{3/2}-8\,^2S_{1/2})$ $7944\ (6\,^2P_{3/2}-8\,^2S_{1/2})$ $8521\ (6\,^2S_{1/2}-6\,^2P_{3/2})$ $8943\ (6\,^2S_{1/2}-6\,^2P_{1/2})$	120° C.	Helium line 3888	Boeckner [2]
Hg	$2537\ (6\,^1S_0-6\,^3P_1)$ $1849\ (6\,^1S_0-6\,^1P_1)$	—	Room temp.	Vacuum arc	Wood [33] and others Rump [23]
Cd	$3261\ (5\,^1S_0-5\,^3P_1)$ $2288\ (5\,^1S_0-5\,^1P_1)$	—	150–250° C. 150–200° C.	Vacuum arc Discharge tubes	Terenin [29] and others
Zn	$3076\ (4\,^1S_0-4\,^3P_1)$ $2139\ (4\,^1S_0-4\,^1P_1)$	—	250–400° C. 250–350° C.	Vacuum arc Discharge tubes	Ponomarev and Terenin [20] Soleillet [27], and others
Tl	$3776\ (6\,^2P_{1/2}-7\,^2S_{1/2})$ $2768\ (6\,^2P_{1/2}-6\,^2D_{3/2})$	$5350\ (6\,^2P_{3/2}-7\,^2S_{1/2})$ $3530\ (6\,^2P_{3/2}-6\,^2D_{3/2})$	450–500° C.	Vacuum arc Discharge tubes	Terenin [29]
Pb	$2833\ (6p^2.\,^3P_0-6p7s.\,^3P_1)$	$3640\ (6p^2.\,^3P_1-6p7s.\,^3P_1)$ $4058\ (6p^2.\,^3P_2-6p7s.\,^3P_1)$	—	Vacuum arc	,,
Bi	$2277\ (6p^3.\,^4S_{3/2}-6p^25d.3)$ $3068\ (6p^3.\,^4S_{3/2}-6p^27s.1)$	$4723\ (6p^3.\,^2D_{3/2}-6p^27s.1)$	—	,,	,,
Sb	$2068\ (^4S_{3/2}-\,^4P_{5/2})$ $2176\ (^4S_{3/2}-\,^4P_{3/2})$ $2311\ (^4S_{3/2}-\,^4P_{1/2})$	$2598\ (^2D_{5/2}-\,^4P_{5/2})$ $2671\ (^2D_{3/2}-\,^4P_{3/2})$ $2770\ (^2D_{5/2}-\,^4P_{3/2})$ $2878\ (^2D_{3/2}-\,^4P_{1/2})$	200° C. Superheat vapour to 1100° C.	,,	,,
Mn	$\left.\begin{matrix}4030 \cdot 8\\4030 \cdot 1\end{matrix}\right\}(a\,^6S-z\,^6P)$ $4034 \cdot 5$ $2798\ (a\,^6S-y\,^6P)$	—	—	Manganese spark	Fridrichson [13]

$-10\cdot4$ and $-25\cdot4$ mÅ. These experiments are a good verification of the assumptions underlying the analysis of the h.f.s. of the resonance line of mercury.

Further work on the resonance radiation of the various h.f.s. components of the 2537 line has to do with the measurements on the polarization of resonance radiation and will be discussed in Chap. v. The h.f.s. of the visible triplet has also been investigated in a more complicated type of resonance radiation in which multiple excitation is involved. These experiments will be discussed in Chap. ii.

REFERENCES TO CHAPTER I

[1] Bates, J. R. and Taylor, H. S., *Journ. Amer. Chem. Soc.* **50**, 771 (1928).
[2] Boeckner, C., *Bureau of Stand. Journ. Res.* **5**, 13 (1930).
[3] Bogros, A., *Compt. Rend.* **183**, 124 (1926).
[4] —— *ibid.* **190**, 1185 (1930).
[5] Cario, G. and Lochte-Holtgreven, W., *Z. f. Phys.* **42**, 22 (1927).
[6] Christensen, C. J. and Rollefson, G. K., *Phys. Rev.* **34**, 1154 (1929).
[7] —— —— *ibid.* **34**, 1157 (1929).
[8] Dunoyer, L. and Wood, R. W., *Phil. Mag.* **27**, 1025 (1914).
[9] Dunoyer, L., *Le Radium*, **10**, 400 (1913).
[10] —— *Compt. Rend.* **178**, 1475 (1924).
[11] Ellett, A., *Journ. Opt. Soc. Amer.* **10**, 427 (1925).
[12] Ellett, A. and MacNair, W. A., *Phys. Rev.* **31**, 180 (1928).
[13] Fridrichson, J., *Z. f. Phys.* **64**, 43 (1930); **68**, 550 (1931).
[14] Houtermans, F. G., *ibid.* **76**, 474 (1932).
[15] Kunze, P., *Ann. d. Phys.* **5**, 793 (1930).
[16] Mitchell, A. C. G., *Journ. Frankl. Inst.* **212**, 305 (1931).
[17] Mrozowski, S., *Bull. Acad. Pol.* (1930 and 1931).
[18] —— *Z. f. Phys.* **78**, 826 (1932).
[19] Pirani, Osram Lamp Works, Germany.
[20] Ponomarev, N. and Terenin, A., *Z. f. Phys.* **37**, 95 (1926).
[21] Puccianti, L., *Accad. Lincei Atti*, **13**, 430 (1904).
[22] Rump, W., *Z. f. Phys.* **29**, 196 (1924).
[23] —— *ibid.* **31**, 901 (1925).
[24] Schein, M., *Helv. Phys. Acta*, Vol. 2, Supp. 2 (1929).
[25] Schnettler, O., *Z. f. Phys.* **65**, 55 (1930).
[26] Schüler, H., *ibid.* **35**, 323 (1926); **59**, 149 (1930).
[27] Soleillet, P., *Compt. Rend.* **184**, 149 (1927).
[28] Strutt, R. J., *Proc. Roy. Soc.* **91**, 511 (1915); **96**, 272 (1919).
[29] Terenin, A., *Z. f. Phys.* **31**, 26 (1925); **37**, 98 (1926).
[30] Thomas, A. R., *Phys. Rev.* **35**, 1253 (1930).
[30a] Weiss, C., *Ann. d. Phys.* **1**, 565 (1929).
[31] Wiedemann, E. and Schmidt, G. C., *Wied. Ann.* **57**, 447 (1896).
[32] Winans, J. G., *Proc. Nat. Acad. Sci.* **11**, 738 (1925).

[33] Wood, R. W., *Researches in Physical Optics*, I and II, Columbia University Press, New York (1913 and 1919).

[34] —— *Phil. Mag.* **3**, 128 (1902); **6**, 362 (1903).

[35] —— *ibid.* **10**, 513 (1905).

[36] —— *ibid.* **18**, 187 (1909).

[37] —— *ibid.* **23**, 680 (1912); *Proc. Phys. Soc.* (London), **26**, 185 (1914).

[38] —— *Phil. Mag.* **44**, 1105 (1922).

[39] Wood, R. W. and Dunoyer, L., *ibid.* **27**, 1018 (1914).

[40] Wood, R. W. and Kimura, M., *ibid.* **32**, 329 (1916).

[41] Wood, R. W. and Mohler, F. L., *Phys. Rev.* **11**, 70 (1918).

[42] Zehden, *Z. f. Phys.* **86** (1933).

CHAPTER II

PHYSICAL AND CHEMICAL EFFECTS CONNECTED WITH RESONANCE RADIATION

1. STEPWISE RADIATION

1 *a*. MERCURY. In earlier paragraphs of this book various simple types of fluorescence have been discussed in which the process giving rise to the fluorescence was due to absorption by normal atoms of the gas. It remains now to consider a type of fluorescence in which the absorption of light by *excited* atoms plays a dominant rôle. This type of fluorescence was first investigated by Füchtbauer [27], who believed that, if mercury vapour were radiated by intense enough light from a mercury arc, other lines of the mercury spectrum besides the two resonance lines 2537 and 1849 would appear.

The apparatus used by Füchtbauer consisted of a quartz resonance tube entirely surrounded by a quartz mercury discharge tube. Precautions were taken to cool the discharge so that sharp lines would be obtained, and a system of liquid mercury reflectors was used to increase the illumination from the discharge. The pressure of the mercury vapour in the resonance tube could be controlled by regulating the temperature of a side tube containing liquid mercury. His experiments showed that with 10–20 amperes current in the arc, and the side tube at 35° C., considerable fluorescence was observed, consisting of practically all of the stronger lines of the mercury spectrum of wave-length longer than 2537. On the other hand, if the side tube was kept in solid CO_2, only a small amount of scattered light could be seen from the tube. On placing a thin glass tube, transparent to all radiation longer than 2537, between the resonance vessel and the exciting light, no fluorescence could be seen. This showed conclusively that the absorption of the resonance line was the first step in the process producing the fluorescence. Füchtbauer supposed that excited mercury atoms reaching the $6\,^3P_1$ state by absorption of light could then absorb other lines from the arc, reaching higher states, and emit various frequencies as a result of this excitation.

It seemed probable, therefore, that at least two quanta of light must have been absorbed successively by the mercury vapour before re-radiation occurred. The experiments of Füchtbauer have since been repeated, and his assumption of "stepwise excitation" has been accepted. For want of a better name this type of fluorescence has since come to be called *Stepwise Radiation*.

Füchtbauer's original experiments were extended in a series of investigations by Terenin[66], by Wood and by Gaviola. The type of apparatus used in most of these experiments is shown in Fig. 14. A quartz cell R containing mercury vapour,

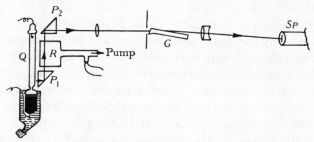

Fig. 14. Apparatus of Wood and Gaviola.

and at times other foreign gases as well, is radiated with light from a water-cooled quartz mercury arc Q. The fluorescence is observed with the help of a right-angled prism P_2 and a spectrograph. When the absorption of the vapour is to be investigated, part of the light from the arc is sent through the quartz cell with the help of the right-angled prism P_1, and the radiation examined for reversal by means of the Lummer-Gehrcke plate G. Sometimes two mercury arcs with various filter combinations were used to study the effect of excitation by several combinations of mercury lines. Furthermore, provision was made to admit several gases at known pressures into the quartz cell. In order to discuss these experiments intelligently it will be necessary to consider the energy-level diagram for mercury given in Fig. 1.

Füchtbauer's experiments showed that a necessary step in

the excitation process was the absorption of the resonance line 2537, thereby raising atoms to the $6\,^3P_1$ state. He did not actually show, however, that the next step in the process was the absorption of a line from the arc which ends on the $6\,^3P_1$ state, although he inferred that such a process did occur, and his explanation is the only possible one based on modern theory. Wood's[75] first experiments consisted in determining which lines appear in fluorescence when the mercury vapour is excited by only certain groups of lines from the arc. In one case he used two mercury arcs, one water-cooled giving an unreversed resonance line and the other run hot so that the core of the resonance line was removed by self-reversal. The light from the water-cooled arc was filtered through a bromine filter, which passed only the lines 2537, 2967 and 3125–3131. Under these conditions no visible radiation could be seen in the resonance tube, but only the ultra-violet lines 3654 and 3663. On lighting the second (hot) arc various visible lines were observed, notably the visible triplet 4047, 4358 and 5461. When the light from the hot arc was filtered through cobalt glass, transmitting only the blue, the green line 5461 still persisted. These experiments show the definite steps in the excitation of the visible triplet, viz. by absorption of 2537 from the water-cooled arc atoms are raised to the $6\,^3P_1$ state, and by the absorption of 4358 from the hot arc the $7\,^3S_1$ state is reached from which the triplet is radiated. Thus one may see that the excitation occurs in two definite steps.

As a further example of the ramifications of this process it is of interest to consider some experiments in which the exciting light was filtered through bromine vapour, which transmits the lines 2537, 2967 and 3125–3131. The visible light emitted from the resonance tube under *strong illumination* is yellow and a spectral photograph shows several ultra-violet lines. The following processes occur: The absorption of 2537 raises atoms to the $6\,^3P_1$ state, some of which drop back to the $6\,^1S_0$ state under emission of 2537, while others absorb lines from the arc terminating on $6\,^3P_1$. Atoms in the $6\,^3P_1$ state may absorb 3131·5 and 3125 and reach the $6\,^3D_1$ and $6\,^3D_2$ states, from which they may return to lower states with emission of 5770 (to $6\,^1P_1$),

3654 (to $6\,^3P_2$) and 3131·5 and 3125. By absorption of the inter-combination line 3131·8 by atoms in the $6\,^3P_1$ state, the $6\,^1D_2$ state may be reached, followed by a subsequent re-emission of 2967·5, 3131·8 and 3663·2. It is to be noted that the line 3650 does not appear under these circumstances. One might expect this line to appear since some atoms land in the metastable $6\,^3P_2$ state by re-emission from the ^3D levels, so that absorption of 3650 and its subsequent re-emission might occur. This takes place to only a small extent. Excitation by unfiltered light, on the other hand, greatly enhances 3650. This is due to the fact that many atoms land in the $6\,^3P_2$ state as a result of the emission of 5461, and the subsequent absorption and emission of 3650 occurs.

1 *b*. EFFECT OF ADMIXTURE OF FOREIGN GASES. The effect on the fluorescence of the admixture of several foreign gases at various pressures was also studied in detail. It was found at the outset that helium, nitrogen, carbon monoxide and argon had a marked effect on the relative intensity of the fluorescent lines. Of the four gases nitrogen was the most thoroughly investigated. The most striking example of the effect of nitrogen is shown by the following observation of the relative intensities of the visible triplet. With no nitrogen in the tube the relative intensities of the lines were 4:2:1 (5461, 4358, 4047), whereas with a few millimetres nitrogen pressure the relative intensity changed to 128:32:4. It will be seen that the intensity of all the lines is increased, but that of 5461 is relatively much more increased than that of 4047. The fact that the intensity of all lines is increased by nitrogen is due to the transfer of $6\,^3P_1$ atoms to the metastable $6\,^3P_0$ state by collision with a nitrogen molecule. In the process, known as a collision of the second kind, the mercury atom loses 0·2 volt energy and the nitrogen gains an equal amount as vibrational energy, no energy being lost as radiation. Such metastable atoms have a much better chance of absorbing a quantum of radiation from the exciting source than have atoms in the $6\,^3P_1$ state, since these metastable atoms have a longer mean life and consequently a larger chance of being in a state to absorb a quantum of 4047. Due to

the long mean life of the $6\,^3P_0$ state it is easy to see why the stepwise lines are increased in intensity. The existence of the long life of the $6\,^3P_0$ state was recognized at about the same time through experiments on sensitized fluorescence (see Chap. II, § 2).

Wood demonstrated conclusively that the increase in intensity of the visible triplet in the presence of nitrogen was due to the production of a large number of atoms in the $6\,^3P_0$ state. To do this he sent part of the light from the arc through the absorption cell by means of the two prisms P_1 and P_2 of Fig. 14 and measured the absorption of the various mercury lines with the help of the Lummer-Gehrcke plate. His results showed that, with mercury vapour alone, none of the lines of the triplet showed any measurable absorption, while with a few milli-metres of nitrogen present the line 4047 was strongly reversed. In later experiments, Gaviola[28] measured the reversal of 4047 as a function of the pressure of the foreign gases nitrogen, carbon monoxide and water vapour, and found varying degrees of self-reversal depending on the nature of the gas and its pressure. Other lines, namely 2967, 5461 and 3650, should also show some absorption in the presence of nitrogen. Of these, 3650 and 5461 are the only lines which have been extensively studied. The line 3650 showed a small amount of reversal, whereas 5461 showed none, indicating that the number of atoms in the $6\,^3P_2$ state is small.

That certain lines show more absorption than others explains the fact that addition of nitrogen increased the intensity of the 5461 line 32 times while it only increased the intensity of the 4047 line four times, a considerable quantity of the light from 4047 being lost by self-absorption. In general, the intensity of practically all the fluorescent lines is changed by the addition of nitrogen, carbon monoxide or water vapour. The exact rela-tion of the intensities of the fluorescent lines to the foreign gas pressure is a more complicated matter and will be treated presently.

As we have seen, most of the lines appearing as stepwise radiation are the result of absorption of two quanta of radia-tion. The emission of the line 3650 is brought about by the absorption of three light quanta. With mercury vapour *in vacuo* normal atoms absorb 2537, are brought to the $6\,^3P_1$ state where they absorb 4358 and are raised to the $7\,^3S_1$ state; subsequent emission of 5461 brings the atom to the $6\,^3P_2$ state, where it absorbs 3650 and is in a position to re-emit this line. The intensity of the fluorescent line 3650 must be proportional to the product of the intensities of the three arc lines producing it (2537, 4358, 3650). If the relative intensity of these lines in the arc is constant, the intensity of the 3650 line in fluorescence should be proportional to the cube of the light intensity of the arc. This effect was shown by Wood and Gaviola[76] as follows: Mercury vapour, in the absence of any foreign gas, was radiated by the total light from a mercury arc. A series of wire screens, which cut down the total illumination by known amounts, was placed between the exciting lamp and the resonance tube. Observations were made on 3650 and 3654. The fact that 3650 is proportional to the third power whereas 3654 is proportional to the second power of the exciting light intensity is shown by the following table.

TABLE IV

Line 3650		Line 3654	
Change	$\sqrt[3]{\text{Change}}$	Change	$\sqrt[2]{\text{Change}}$
150 times	5·3	30 times	5·5
1200 ,,	10·6	120 ,,	11·0
240 ,,	6·2	40 ,,	6·3
150 ,,	5·3	30 ,,	5·5
240 ,,	6·2	40 ,,	6·3
400 ,,	7·4	50 ,,	7·1
480 ,,	7·8	60 ,,	7·7

An approximate theoretical treatment of this effect has been given by Gaviola[28].

Gaviola has made a careful experimental and theoretical investigation of the effect of foreign gases on the intensity of the fluorescent lines. He used the apparatus shown in Fig. 14 and measured the absorption of the line 4047 as a function of the distance away from the wall of the tube through which the exciting light enters and of the pressure of the foreign gas.

With the beam of light, whose absorption is to be measured, traversing the tube at a fixed distance (not given) from the front wall of the tube, he investigated the structure of the absorption line with different pressures of CO, H_2O and N_2. A clear reversal of the main component of 4047 was to be seen with 0·015 mm. CO, the reversal increasing up to 0·2 mm. pressure and then decreasing again such that at pressures above 4 mm. no more reversal could be seen. Water vapour causes reversal at 0·05 mm. pressure and the reversal increases steadily until at 2 mm. pressure the whole line is absorbed. With N_2, on the other hand, no definite reversal is shown until a pressure of 0·5 mm. is reached; below these pressures, however, a diffuse broadening of the line occurs.

In order to test the absorption of 4047 as a function of depth, he placed two slits across the resonance tube in such a way as to allow the beam of 4047 to traverse the resonance tube at distances of 2 cm. and 5 cm. from the illuminated wall, respectively. At 2 cm. from the wall and with from 0·2 to 0·5 mm. water vapour pressure, self-reversal is clearly seen, but disappears again if the pressure reaches 2 mm. At 5 cm. depth no self-reversal is seen at any time. These facts show that diffusion of metastable atoms does actually take place at low water vapour pressure and that the mean life of the metastable atoms is long enough to allow them to diffuse 2 cm. but not 5 cm. The same experiments were performed with N_2 with the result that no reversal could be found at 2 cm. depth with any nitrogen pressure. This result is to be expected, because Gaviola had shown that reversal does not set in until at a pressure of 0·5 mm., which is already too high to allow diffusion to take place.

From an approximate consideration of the diffusion of metastable atoms and the absorption coefficient of various lines in the excited vapour, Gaviola was able to show that the intensity of the several fluorescent lines should vary with the depth from the illuminated window at which these lines are observed. After removing the slit system and Lummer-Gehrcke plate, he focused an image of the resonance tube on the slit of the spectrograph and was able to show that the intensity of the fluorescent lines changed as a function of distance from the illuminated wall and the conditions of excitation. Certain lines, notably 4358, persisted to considerable distances away from the wall with pressures up to 2 mm. of N_2, whereas others such as 4047 persisted to large distances at low N_2 pressures and to only short distances at higher pressures. The results of Gaviola's experiments and calculations explain qualitatively the experimental fact that certain fluorescent lines are enhanced by the addition of N_2 while others are somewhat weakened, although addition of nitrogen causes an increase of the number of metastable $6\,^3P_0$ mercury atoms present.

Another series of experiments showing quantitatively the absorption of the 4047 line by the excited mercury vapour in the resonance tube as a function of foreign gas pressure was performed by Klumb and Pringsheim [37]. Their apparatus was somewhat similar to that previously described. Their resonance vessel was a quartz tube with plane ends and was illuminated from the side by a water-cooled quartz mercury arc. Light from a similar arc was projected by a suitable lens system along the axis of the tube and, after passing through a monochromatic illuminator set to transmit the line 4047, was received by a photoelectric cell. Observations could be made as follows: (1) light from both arcs cut off (zero point measurement), (2) intensity of 4047 with the exciting light (arc on side of resonance tube) cut off (J_1), and (3) intensity of 4047 with both arcs illuminating the tube (J_2). The relation $E = \dfrac{J_1 - J_2}{J_1}$ is a measure of the absorption of 4047 in the resonance tube. Measurements of the absorption of the line 4047 as a function of N_2 pressure, together with admixtures of other gases, are shown in Fig. 15. It will be noted that the absorption is zero at zero foreign gas pressure, rises sharply to a maximum of 50 per cent. at about 1 mm. pressure, and then stays constant. It should be noted further that the addition of 10^{-4} mm. of H_2 markedly reduces the absorption for a given N_2 pressure. This is to be expected, since H_2 is known to destroy metastable mercury atoms. The curve for the absorption in the presence of water vapour is slightly different, in that the absorption reaches a maximum at 0·5 mm. pressure and then decreases with higher vapour pressure, in agreement with Gaviola's results.

1c. THE APPEARANCE OF THE FORBIDDEN LINE 2656 ($6\,^1S_0$–$6\,^3P_0$). Wood and Gaviola[76] observed that with small quantities of nitrogen or water vapour in the resonance tube the forbidden line 2656 ($6\,^1S_0$–$6\,^3P_0$) appeared. They found that water vapour was more efficient in producing the line than was nitrogen. This is what one would expect, since water vapour is more efficient in producing metastable atoms than is nitrogen.

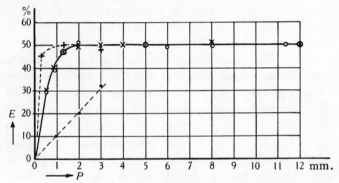

Upper curves: — pure nitrogen; - - - with 10^{-4} mm. H_2;
Lower Curve: – – – with 12 mm. He-Ne.

Fig. 15. Absorption of 4047 as a function of foreign gas pressure.

They also found that the intensity of the forbidden line was proportional to the first power of the intensity of the exciting light. This line, although considered as forbidden by the exclusion principles of the older quantum theory, cannot be said to be strictly forbidden. The new quantum mechanics shows that such lines may have a small but finite probability of occurrence per mercury atom in the $6\,^3P_0$ state. The above-mentioned experiments have increased the number of $6\,^3P_0$ atoms present to such an extent that the "forbidden" line appears.

1d. THE HYPERFINE STRUCTURE OF STEPWISE RADIATION. As has already been mentioned, Collins[16] and Mrozowski[53] have investigated the h.f.s. of the visible triplet lines in mercury obtained by the method of stepwise excitation. Collins used two mercury arcs to excite fluorescence in a tube containing (*A*) mercury vapour alone, and (*B*) mercury vapour together with a few millimetres of nitrogen. Observation of the fluorescence with a Lummer plate and spectrograph showed that only the strong central component of the 5461 line appeared as fluorescence when mercury vapour alone was in the tube. With a few millimetres of nitrogen present, however, the fluorescent line showed two components, the 0·0 and the $-23·5\,\text{m\AA}$. Mrozowski, exciting with filtered 2537 light together with light from a second arc giving only the visible lines, found only the main component present when nitrogen was present. The two experiments seem contradictory, but this contradiction may be only an apparent one, since the intensity of the exciting sources in the two cases may have been quite different. The main point of the experiment is that not all of the components of 5461 (there are some twelve of them) are seen in fluorescence but only the strongest one. A similar result was found with the other two lines of the triplet 4047 and 4358 in fluorescence. Collins reported several components of each line both with and without nitrogen in the tube, and differences in the number of components depending on whether nitrogen is present or not. Mrozowski, on the other hand, found both lines to consist of only the central component.

Mrozowski further investigated the structure of the fluorescent lines when they were excited by different h.f.s. components of the resonance line, together with all the h.f.s. components of 4047. He found the rather surprising result that only the central component of the triplet lines appeared, no matter whether he excited with the $-25·4\,\text{m\AA}$. component of the 2537 line (containing only lines from the isotopes of odd atomic weight) or with components containing only lines from the isotopes of even atomic weight. He explains this by assuming that metastable mercury atoms of a given isotopic kind can excite normal mercury atoms of a different isotopic kind to the metastable level by collision of the second kind. He further showed that the relative intensity of the triplet lines in fluorescence was independent of the h.f.s. components of 2537 used in the exciting beam.

One should remark at this point, in regard to Collins's experiments, that changes in the number and relative intensity of h.f.s. components of the visible triplet lines seen in stepwise fluorescence are to be expected when nitrogen is introduced into the tube. The factors governing the change are rather complicated, so that no exact theoretical prediction has yet been made. As to the result when no nitrogen is present, the intensity of the h.f.s. components of a given stepwise line will depend on the structure of the 4358 line, whereas, if nitrogen is present, the structure of the line 4047 will govern the excitation. Furthermore, the presence of nitrogen itself adds complications.

1e. CADMIUM AND ZINC. Since cadmium and zinc show similar spectra to mercury, differing only in separation of energy levels, Bender[4] thought it worth while to investigate the stepwise radiation exhibited by these elements. His experimental arrangement consisted of a resonance tube, of the usual shape, surrounded by a coil of quartz tubing, through which passed a high potential discharge in hydrogen and cadmium (or zinc). The resonance tube was temperature controlled and the vapour pressure of the cadmium kept constant at 0·008 mm. The fluorescence was observed end on, and precautions were taken to avoid scattered light from the quartz surfaces of the tube. It was found that an intense bluish-green fluorescence was observable when the resonance tube was excited by the Cd-H_2 discharge, but that the tube emitted no light if cadmium was not present in the discharge. The fluorescence, with cadmium vapour alone in the resonance tube, contained all the strong lines in the cadmium spectrum (see Fig. 16 for spectrum of cadmium) except 2288.

The effect of the addition of the gases nitrogen and carbon monoxide on the stepwise radiation of cadmium was quite different from the effects observed with mercury. The addition of 0·01 mm. of nitrogen produced an observable increase in the intensity of the 3404 line, which attained a maximum of intensity at 0·1 mm. nitrogen pressure. At pressures above 0·1 mm. the ratio of intensity of 3404 to the rest of the spectrum remained constant at 2 to 1. A similar effect was produced on this line by carbon monoxide. There appeared to be no

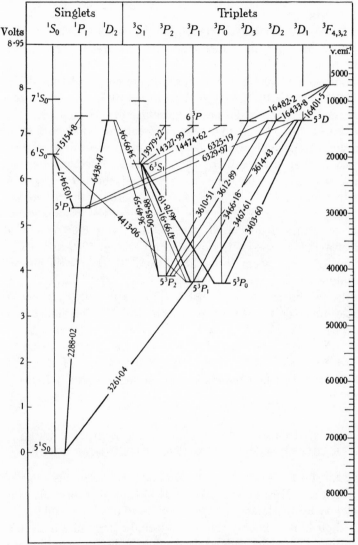

Fig. 16. Energy levels of cadmium.

enhancement of the visible triplet ($6\,^3S_1$–$5\,^3P$) due to the presence of nitrogen or carbon monoxide as was the case with mercury fluorescence, but merely a general quenching of the whole spectrum. The enhancement of the 3404 line is probably due to the formation of metastable $5\,^3P_0$ atoms, as in the case of mercury.

There are certain fundamental differences between the spectra of cadmium and mercury which are important. In the first place the mean life of the $5\,^3P_1$ state of cadmium is about 20 times longer than that of mercury. (Mean life of Cd $5\,^3P_1 = 2 \cdot 5 \times 10^{-6}$, of Hg $6\,^3P_1 = 1 \cdot 08 \times 10^{-7}$.) This accounts for the fact that smaller pressures suffice to give enhancement of the 3404 line than are necessary to enhance analogous lines in the mercury spectrum. The ratio of the pressures necessary is about that of the ratio of the mean lives. Furthermore, the energy difference between the $5\,^3P_1$ and the $5\,^3P_0$ states in cadmium is only $0 \cdot 07$ volt, whereas in mercury the difference is $0 \cdot 218$ volt. The relative kinetic energy of the gas molecules at the temperatures used ($350°$ C.) is about $0 \cdot 08$ volt and corresponds to the energy difference between $5\,^3P_1$ and $5\,^3P_0$. Thus any $5\,^3P_1$ atoms brought to the $5\,^3P_0$ state by collision would have a good chance of being transferred back to the $5\,^3P_1$ state on the next collision. This explains why there is little enhancement of the spectrum by the addition of nitrogen and carbon monoxide. The main effect of these gases is therefore a general quenching of the whole fluorescent spectrum.

In the case of zinc vapour, stepwise radiation was found, but the effect of carbon monoxide and nitrogen was not investigated.

2. PRODUCTION OF SPECTRA BY COLLISION WITH EXCITED ATOMS: SENSITIZED FLUORESCENCE

$2a$. THE PRINCIPLE OF MICROSCOPIC REVERSIBILITY. Franck and Hertz discovered in 1913 that electrons, given a velocity by acceleration through a potential field, would transfer their kinetic energy into the internal energy of a molecule or atom. The atom thus excited might subsequently give up this energy as radiation. Such a process, in which a fast electron collides with a slow-moving atom and which results in the formation of an excited atom and a slow electron, has been termed a *collision of the first kind*.

In order to preserve thermodynamic equilibrium in a mixture of atoms and electrons it is necessary to assume that some kind of reverse process to the one explained above must occur in which fast electrons and unexcited atoms result. Not only must we assume that at equilibrium the overall rate of formation of fast electrons and unexcited atoms must be the same as the overall rate of formation of excited atoms and slow electrons, but we are forced to make the postulate that: "The total number of molecules leaving a given quantum state in unit time shall equal the number arriving in that state in unit time, and also the number leaving by any one particular path shall be equal to the number arriving by the reverse path".* The postulate which entails that each *microscopic* process occurring must be accompanied by an inverse process is called the Principle of Microscopic Reversibility.

Klein and Rosseland [35], making use of this type of reasoning, therefore postulated that if fast electrons could collide with unexcited atoms and produce excited atoms and slow electrons, then the reverse process must occur, namely that excited atoms may collide with slow electrons and produce unexcited atoms and fast electrons. The process must, of course, be unaccompanied by radiation. Such a process has been called a *collision of the second kind*.

2*b*. EFFICIENCY OF COLLISIONS OF THE SECOND KIND BETWEEN ATOMS AND ELECTRONS. Klein and Rosseland made calculations from which they could make some statements as to the efficiency of the processes. In order to carry out the calculations, consider an ensemble of atoms and electrons. The atoms will be considered to have only two states, a lower state 1 of energy E_1 and an upper state 2 of energy E_2. Now the number of atoms in each state at equilibrium is given by

$$n_1 = C p_1 e^{-\frac{E_1}{kT}}; \quad n_2 = C p_2 e^{-\frac{E_2}{kT}} \qquad \ldots\ldots(4),$$

* See R. C. Tolman, *Proc. Nat. Acad. Sci.* 11, 436 (1925), where the above statement and a discussion of the Principle of Microscopic Reversibility are to be found.

where p_i is the statistical weight of the ith state and C is a constant independent of E. Now the number of electrons having energies between E and $E+dE$ is given by the Maxwell-Boltzmann law as

$$\mu(E)\,dE = Ke^{-\frac{E}{kT}}\sqrt{E}\,dE \qquad \ldots\ldots(5),$$

where K is again a constant independent of E. Let us define $S_{12}(E)$ as the probability of a collision of the first kind in such a way that the total number of collisions of the first kind taking place per second is

$$S_{12}(E)\,n_1(E_1)\,\mu(E)\,dE \qquad \ldots\ldots(6).$$

Similarly, for the number of reverse processes per second we have

$$S_{21}(E)\,n_2(E_2)\,\mu(E)\,dE \qquad \ldots\ldots(7).$$

Since Franck and Hertz found that for electron energies less than $E_2 - E_1$ no excitation is possible, it follows that

$$S_{12}(E) = 0 \quad \text{for} \quad E < E_2 - E_1 \qquad \ldots\ldots(8).$$

Consider now the equilibrium of electrons in the energy range dE between E' and $E' + dE$ when $E' < E_2 - E_1$. Electrons can obviously only leave this energy range by collision of the second kind, and the number leaving is given by

$$n_2 S_{21}(E')\,\mu(E')\,dE.$$

The number entering this energy range by collision of the first kind must have originally had energies between

$$E'' = E' + E_2 - E_1 \quad \text{and} \quad E'' + dE.$$

The number is given by

$$n_1 \mu(E'')\,S_{12}(E'')\,dE.$$

At equilibrium, therefore, we have

$$n_2 \mu(E')\,S_{21}(E') = n_2 \mu(E'')\,S_{12}(E'') \qquad \ldots\ldots(9).$$

Remembering that

$$E'' - E' = E_2 - E_1 \qquad \ldots\ldots(10),$$

it follows that

$$S_{21}(E') = \frac{p_1}{p_2}\left(\frac{E''}{E'}\right)^{1/2} S_{12}(E'') \qquad \ldots\ldots(11).$$

A similar argument can be considered for the case in which

$E' > E_2 - E_1$ and leads likewise to (11). One may see from (11) that, since

$$E'' \geqslant E' \text{ and } p_1 \cong p_2, \quad S_{21}(E') \geqslant S_{12}(E''),$$

which means that a collision of the second kind between a slow electron and an excited atom must be very probable. This is to be expected for electrons and atoms, at any rate, since a slow electron will remain in the neighbourhood of an atom longer than a fast one and the probability of energy transfer will therefore be greater. Investigations confirming the above theory have been made by Smyth[65], Latyscheff and Leipunsky[41], Kopfermann and Ladenburg[39], and Mohler[51], and will be discussed further in Chap. IV.

2c. COLLISIONS OF THE SECOND KIND BETWEEN TWO ATOMS. Franck[24] extended the ideas of Klein and Rosseland to include collisions between two atoms or molecules. Thus he supposed that an excited atom might collide with a normal atom or molecule and give up a quantum of energy $(E_2 - E_1)$ to the unexcited atom; the latter might then take up the energy either as translational energy, excitational energy or both, there being no loss of energy by radiation during the process. Such radiationless transfers of energy Franck also called collisions of the second kind.

Many examples of these processes exist, and a list of them will be found in Chap. IV. We are concerned in this chapter with the bearing of such collisions on two important phenomena, namely, sensitized fluorescence and sensitized chemical reactions.

2d. SENSITIZED FLUORESCENCE. Consider a mixture of two kinds of atoms A and B, which for simplicity we shall suppose to have only one excited and one normal state. Let the energy state be represented as in Fig. 17, the excited state of A lying higher than that of B by an amount ΔW. Let this mixture be irradiated by light of frequency ν. If the number of atoms of the kind A is sufficient, there will be a considerable absorption of the frequency ν and consequently some re-emission of the same frequency. If now the number of atoms of the kind B is large enough, so that the time between collisions between A

and B is of the same order of magnitude as the mean life of the excited state of A, then, by collision of the second kind between excited A atoms and B atoms, energy will be transferred to the B atoms, and there will be a subsequent emission of the frequency ν together with the frequency ν_1. The difference in energy ΔW will then appear as relative kinetic energy of A and B. From the laws of conservation of energy and momentum one can calculate what fraction of this kinetic energy is carried by A and B. Suppose the temperature is so

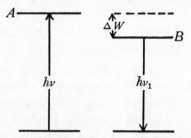

Fig. 17. Illustrating sensitized fluorescence.

low that the energy of thermal agitation is small compared to ΔW. Then from conservation of energy and momentum we have

$$\tfrac{1}{2}m_A v_A{}^2 + \tfrac{1}{2}m_B v_B{}^2 = h\nu - h\nu_1 = \Delta W \quad \ldots\ldots(12),$$

and
$$m_A v_A = m_B v_B,$$

from which it follows that

$$\left.\begin{aligned}
\tfrac{1}{2}m_A v_A{}^2 &= \Delta W \frac{m_B}{m_A + m_B} \\
\tfrac{1}{2}m_B v_B{}^2 &= \Delta W \frac{m_A}{m_A + m_B}
\end{aligned}\right\} \quad \ldots\ldots(13),$$

where v_A and v_B are the final velocities of A and B respectively. From (13) it is easy to see that, if ΔW is large, the atom B will acquire a considerable velocity from the collision, especially if it is light. This fact may be demonstrated by the existence of a Doppler effect on the line of frequency ν_1, which should be broadened by an amount $\Delta \nu$ given by the well-known Doppler equation

$$\frac{\Delta \nu}{\nu} = \frac{v}{c}\cos\phi \quad \ldots\ldots(14).$$

These predictions have all been verified by experiment. Cario and Franck[12, 14] tested the theory by experiments on mercury and thallium vapours. Their experimental arrangement is shown in Fig. 18. A quartz tube Q, containing thallium and mercury vapours, is illuminated by the light of a well-cooled quartz mercury arc lamp. A vapour pressure of thallium of about 2 mm. is obtained by heating a globule of thallium to 800° C. in a side tube contained in the oven O_2. A high pressure of mercury vapour is also obtained by heating mercury to 100° C. in the oven O_3. The furnace O_1 is kept at a temperature above 800° C. to keep thallium from condensing in the tube. Under these conditions of vapour pressure very bright fluorescence takes place in a very small layer close to the front of the

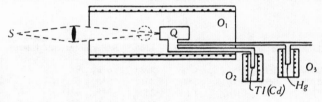

Fig. 18. Apparatus of Cario and Franck.

resonance tube*. A spectrogram of the fluorescence shows a number of thallium lines in addition to the 2537 line of mercury. In fact the intensity of the green thallium line 5350 is so marked that the experiment can be demonstrated before a class. When the mercury arc is run without cooling, so that the 2537 line is reversed, the thallium fluorescence disappears completely. On freezing out the mercury vapour in the tube a similar disappearance of the fluorescence is noted.

A list of some of the thallium lines occurring as fluorescence is given in Table V together with their classification.

It will be seen that three of these lines come from energy levels which lie higher than 4·9 volts, the excitation energy of mercury. The explanation probably lies in the fact that kinetic

* Such high vapour pressures of mercury obviously do not have to be employed to obtain a measurable fluorescence. For example, Mitchell (*Journ. Frankl. Inst.* **209**, 747 (1930)), was able to obtain sensitized fluorescence of cadmium and thallium when the mercury vapour pressure corresponded to that of room temperature.

energy, obtained through high temperatures, can co-operate with excitational energy to excite higher states. However, it is difficult to prove this statement conclusively by experiments on thallium on account of the low-lying metastable $6\,^2P_{3/2}$ level. The relative intensities of the lines 2768 and 3776 may be taken as a test of Eqs. (13) and (14). It was observed that these lines did not occur with the same relative intensity with which they occur in the arc. For example, the line 3776 was very strong whereas 2768 was weak. The reason for this, given by the authors, is that the 2768 line comes from the $6\,^2D_{3/2}$ level, lying only 0·4 volt below the excitation voltage of mercury. The emitted line, therefore, is narrow and is absorbed on passing through the thallium vapour in the tube. The 3776

TABLE V

Line	Series notation	Energy	Line	Series notation	Energy
2580	$6\,^2P_{1/2}-8\,^2S_{1/2}$	4·78	3230	$6\,^2P_{3/2}-8\,^2S_{1/2}$	4·78
2709	$6\,^2P_{3/2}-8\,^2D_{5/2}$	5·60	3519	$6\,^2P_{3/2}-6\,^2D_{5/2}$	4·51
2768	$6\,^2P_{1/2}-6\,^2D_{3/2}$	4·45	3530	$6\,^2P_{3/2}-6\,^2D_{3/2}$	4·45
2826	$6\,^2P_{3/2}-9\,^2S_{1/2}$	5·36	3776	$6\,^2P_{1/2}-7\,^2S_{1/2}$	3·27
2918	$6\,^2P_{3/2}-7\,^2D_{5/2}$	5·24	5350	$6\,^2P_{3/2}-7\,^2S_{1/2}$	3·27

line, on the other hand, comes from the $7\,^2S_{1/2}$ level, lying 1·6 volts below the $6\,^3P_1$ level of mercury, and is consequently broadened by Doppler effect and is not weakened by absorption.

A quantitative confirmation of this effect was obtained by Rasetti[61], who investigated the sensitized fluorescence of sodium. The energy excess of excited mercury above that necessary to excite the D lines is 2·8 volts which, on applying Eq. (11), would give the excited sodium atoms a velocity of $4·3 \times 10^5$ cm./sec. The distribution of the velocities of the excited sodium atoms is uniform as regards direction in space, but is not a Maxwellian one, since all the molecules have the same velocity. The emission line should be broadened by the amount $\Delta\lambda = \dfrac{2\lambda v}{c}$, and the intensity distribution should be uniform. Taking v as $4·3 \times 10^5$, $\Delta\lambda$ is equal to 0·17 Å. Using a 40-plate echelon grating Rasetti measured the breadth of both D_1 and D_2. A mean of six observations gave $\Delta\lambda = 0.16$ Å. in remarkable agreement with theory. The reader should note, however, a further discussion of these results in § 2f.

In order to make an unambiguous test as to whether translational energy and excitational energy can co-operate to excite higher quantum states, the sensitized fluorescence of cadmium was studied. The advantage of cadmium (see Fig. 16) lies in the fact that it has no low-lying metastable states which might be excited by collision of the first kind with other normal atoms having thermal velocities. The experiment was performed with the resonance tube at 800° C., and it was found that not only did the line 3261 appear but also the visible triplet from the level $6\,^3S_1$, having an excitational energy of 6.3 volts. When the experiment was repeated with the resonance tube at 400° C., the visible triplet did not appear. In order to be sure that this effect was not a result of stepwise radiation in mercury, the experiments were repeated using a monochromatic illuminator passing only the line 2537, with the same results. It is therefore clear that in order to excite the $6\,^3S_1$ state of cadmium, excitational energy of mercury (4.9 volts) must co-operate with translational energy from temperature motion.

A similar experiment on a mixture of zinc and mercury vapour was carried out by Winans [74], who found still other effects than those reported by Cario and Franck. The experiments were made in a sealed-off quartz tube containing mercury vapour at 0.28 mm. and zinc vapour at 16 mm. pressure. The tube was kept at 720° C. Light filters were used to give excitation by various combinations of mercury lines. It was found that when the mixture was irradiated by the full spectrum of the water-cooled mercury arc (1849–7000 Å.) the lines given in Table VI appeared.

It will be noted that the lines coming from states with energies as high as 7.74 volts appeared. When the mixture was excited by wave-lengths from 3200–7000 Å., no zinc lines appeared, which shows that lines in the region below 3200 Å. are necessary. With incident radiation of from 2300–7000 Å. all lines appeared except 2138, the intensity of the sharp triplet being much reduced, however, by the omission of the wave-lengths between 1849 and 2300. When the exciting light consisted of only 2537 and 4358 no lines except 3075 appeared. If the exciting light contained all wave-lengths from 1849 to 7000 without the core of 2537, only the sharp triplet appeared. The same was true when the exciting light contained wave-lengths from 1849–2000 with light from the mercury arc (either hot or cold) or from the aluminium spark. Finally, with wave-lengths between 1950 and 2000 no sensitized fluorescence occurred.

The results are to be explained as follows. Since the sharp triplet only appeared brightly when the exciting light con-

TABLE VI

Wave-length	Series notation	Energy necessary to excite
4810	$4\,^3P_2 - 5\,^3S_1$	
4722	$4\,^3P_1 - 5\,^3S_1$	6·62
4680	$4\,^3P_0 - 5\,^3S_1$	
3344	$4\,^3P_2 - 4\,^3D_3$	
3302	$4\,^3P_1 - 4\,^3D_1$	7·74
3282	$4\,^3P_0 - 4\,^3D_1$	
3075	$4\,^1S_0 - 4\,^3P_1$	4·01 Resonance
2138	$4\,^1S_0 - 4\,^1P_1$	5·76 lines

tained wave-lengths near 1849, it is certain that the $6\,^1P_1$ state of mercury must have been excited. In general, however, 1849 is highly reversed in a mercury arc. Since the $6\,^1P_1$ state of mercury was excited by a hot arc and aluminium spark it is to be inferred that the absorption line at 1849 was highly broadened due to pressure, or perhaps even molecule formation. The fact that the diffuse zinc triplet did not appear when only 2537 and 4358 were in the exciting light shows that these lines are probably brought about by collision with some higher excited state of mercury formed by stepwise radiation. A small percentage of the intensity of the sharp triplet is also probably due to collision with mercury atoms in higher states due to stepwise radiation.

The results of this experiment are not to be taken as in disagreement with those of Cario and Franck on cadmium, since the amount of energy necessary to reach the $5\,^3S_1$ state of zinc is greater than that necessary in cadmium, and the temperatures employed here were not so high.

The sensitized fluorescence of many other metals has been investigated as an aid to finding their energy levels. The following metals have been studied: thallium [12], silver [12], cadmium [14], sodium [7], lead, bismuth [38], zinc [74], indium, arsenic and antimony [21] (the two last giving negative results).

2e. EFFECT OF METASTABLE ATOMS. In view of the fact that Wood and Cario had shown that the addition of a foreign gas quenched mercury resonance radiation, Donat [21], and later Loria [43], made investigations of the effect of added gases on the sensitized fluorescence of thallium. The gases argon, nitrogen and hydrogen were used. It was to be expected that since these gases were known to remove mercury atoms from the $6\,^3P_1$ state (quenching), the sensitized fluorescence would also be quenched as a result. However, on performing the experiment with argon or nitrogen, the intensity of the thallium lines was found to increase. This experiment can be explained in the following way. Mercury atoms are raised to the $6\,^3P_1$ state by absorption of 2537; collision with argon or nitrogen then brings them to the metastable $6\,^3P_0$. These atoms have a long mean life, and also appear to be able to survive many collisions with nitrogen molecules and argon atoms without losing their activation. They are therefore able to remain activated until they make a collision with a thallium atom. This causes a corresponding increase in the intensity of the sensitized fluorescence, since, without nitrogen or argon, a considerable fraction of the normal mercury atoms would lose their energy by radiation before collision. These experiments, therefore, are in agreement with and supplement those of Wood on stepwise radiation. Donat measured the change of intensity of the mercury 2537 line and the thallium lines as a function of nitrogen and argon pressure. He found qualitatively that the intensity of 2537 which was lost owing to quenching was gained by the thallium lines. Loria and Donat found that there was a certain pressure of nitrogen or argon which gave the greatest increase in intensity of the thallium lines, the optimum pressure for argon being greater than that for nitrogen, as one would expect from quenching data. It will be seen that the existence of an optimum pressure for sensitized fluorescence is in agreement with experiments on stepwise radiation. Hydrogen, on the other hand, showed a quenching effect both on the 2537 line and on the thallium lines. This shows that collisions with hydrogen always result in the formation of mercury atoms in the normal state. Sensitized fluorescence may therefore be used as a criterion for telling whether a given gas quenches excited mercury atoms to the normal ($6\,^1S_0$) or metastable ($6\,^3P_0$) state.

Orthmann and Pringsheim [59] showed that collisions between excited and normal mercury atoms lead to the production of metastable atoms. They repeated the Cario and Franck experiment keeping the thallium pressure at 2×10^{-2} mm. ($610°$ C.), whereas the mercury vapour pressure was gradually increased. They noticed that, as the pressure increased up to one atmosphere, the thallium lines lost none of their original intensity, whereas the mercury radiation was completely quenched.

$2f$. EFFICIENCY OF COLLISIONS OF THE SECOND KIND BETWEEN ATOMS. We have mentioned that a collision of the second kind between one atom and one electron is most efficient when the electron has a small velocity. When a collision of the second kind between two atoms occurs, a similar relation holds; viz. the collision will be most efficient when the least energy is converted into kinetic energy [23]. That this effect is to be expected theoretically was shown by Nordheim [54] and by Carelli [11]. Later developments of the wave mechanics [34] have shown that if two atoms have energy levels lying near together, a "quantum mechanical resonance" effect takes place between them. As an illustration consider an atom A which is in an excited state having an energy of 5 volts, say. This atom makes a collision, while still excited, with an unexcited atom B which has two energy levels, one at $4·9$ volts, the other at $4·0$. The quantum theory says that a very strong interaction (resonance) will take place between the atom A and the $4·9$

volt energy level of B, which will lead to a very great probability of the 4·9 volt level of B being excited. The probability of the 4·0 volt level being excited is, on the other hand, much smaller.

Several attempts have been made to test these theories experimentally, but most of the earlier ones led to no definite conclusions, due to various complicating factors. Beutler and Josephy[7], however, succeeded in showing this effect very beautifully by experiments on the sensitized fluorescence of

TABLE VII

Na term	Emission line (Å.)		Energy (volts)	Energy difference (volts)		Hg term
				6^3P_1	6^3P_0	
$4\,^2$D	5688	5683	4·259	− 0·601	− 0·383	
$5\,^2$P	2853	—	4·322	− 0·538	− 0·320	
$6\,^2$S	5154	5149	4·485	− 0·375	− 0·157	
$5\,^2$D	4983	4979	4·567	− 0·293	− 0·075	
$6\,^2$P	2680	—	4·599	− 0·261	− 0·043	
	(2656)		(4·642)	(− 0·218)		6^3P_0
$7\,^2$S	4752	4748	4·687	− 0·173	+ 0·045	
$6\,^2$D	4669	4665	4·734	− 0·126	+ 0·092	
$7\,^2$P	2594	—	4·751	− 0·109	+ 0·109	
$8\,^2$S	4545	4543	4·805	− 0·055	+ 0·163	
$7\,^2$D	4497	4494	4·834	− 0·026	+ 0·192	
$8\,^2$P	2544	—	4·846	− 0·014	+ 0·204	
	(2537)		(4·860)		(+ 0·218)	6^3P_1
$9\,^2$S	4223	4420	4·880	+ 0·020	+ 0·238	
$8\,^2$D	4393	4390	4·899	+ 0·039	+ 0·257	
$9\,^2$P	2512	—	4·907	+ 0·047	+ 0·265	
$10\,^2$S	4345	4341	4·930	+ 0·070	+ 0·288	
$9\,^2$D	4320	4316	4·943	+ 0·083	+ 0·301	

sodium vapour. Sodium has many energy levels lying between 0 and 5·0 volts, several of which lie very close to 4·860 volts (6^3P_1 state of Hg) and 4·642 volts (6^3P_0 state), as is shown in Table VII. By exciting with 2537, they observed the intensities of various sodium series appearing in sensitized fluorescence. One would expect, if the series were excited in the usual way, that the intensity of the higher numbers of the series (lines coming from states with large n) would decrease monotonically with n. On the contrary, Beutler and Josephy found that within the series $3\,^2$P–$n\,^2$S the line $3\,^2$P–$9\,^2$S was by far the strongest line observed. The state from which this line comes lies within 0·020 volt of the $6\,^3P_1$ state of mercury. The state $8\,^2P$ lies even closer (0·014 volt), but this could not be observed as it gives rise to the line 2544, which lies too close to the strong mercury line 2537 to allow of intensity comparison. Fig. 19 shows the result of the measurements. In constructing this figure the intensity of the various lines was corrected for sensitivity of the plate, divided by ν (the frequency) and g (the

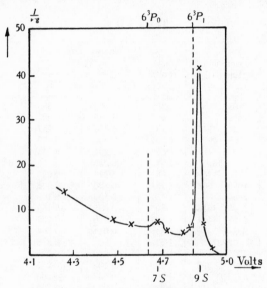

Fig. 19. Intensity relations in sensitized fluorescence of sodium.
(Beutler and Josephy.)

statistical weight of the upper state) to place them on an equal footing for comparison. The ordinates give the probability of excitation of a given state and the abscissae the energy of the states. It will be seen at once that the $9\,^2S$ state, lying close to $6\,^3P_1$, shows an extremely high probability of excitation. A small maximum also occurs at the $7\,^2S$ state, lying close to the metastable $6\,^3P_0$ level. Since the experiments were performed at rather high mercury pressures some metastable atoms were present. One should remark that since the lines $3\,^2P\text{--}n\,^2S$ are not resonance lines the results are wholly free from complications arising from absorption. A similar result was obtained by Webb and Wang[73] by mixing sodium vapour with excited atoms from the arc stream of a mercury arc. The results are in general agreement with those of Beutler and Josephy, although the experimental conditions were not as clean cut as theirs.

A remark about Rasetti's experiments (p. 62) is now necessary. He observed that the D lines excited by sensitized fluorescence were very broad, but did not concern himself with the higher series lines emitted. Since D lines will be emitted as a result of excitation of the higher series members as well as by direct excitation of the $3\,^2$P states of sodium, not much weight can be attached to the quantitative results of Rasetti.

$2g$. CONSERVATION OF SPIN ANGULAR MOMENTUM IN COLLISIONS OF THE SECOND KIND. The new quantum mechanics* predicts another effect which has a bearing on the efficiency of collisions of the second kind. This effect depends on the electron configurations of two atoms undergoing a collision. Let us consider the particular case of atoms having an even number of electrons. Each electron in the atom may be considered as spinning, and the resultant spin, s, of all the electrons in a given state determines the multiplicity of this state (multiplicity $= 2s + 1$). If two electrons spin with their spin vectors antiparallel to each other they may be said to be paired, and the resultant spin is zero. A state in which all of the electrons are paired will have zero resultant spin. If, on the other hand, two electrons spin with their spin vectors parallel, they are not paired. A state in which two of the electrons are unpaired will have a resultant spin of one. If two atoms are about to make a collision of the second kind, Wigner's theorem states that, of all the possible transfers of energy, *that* one will be most likely to occur in which the *total resultant spin*, considered for the two atoms together, remains unchanged. Consider now the case of a krypton atom in the metastable $[4p^5\,(^2\mathrm{P}_{3/2})\,5s]$ state (to be denoted by $^3\mathrm{P}_2$) which may make a collision with a mercury

* E. Wigner, *Gött. Nachr.* 375 (1927).

atom in the normal state ($6\,^1S_0$). We may write the electron configuration for the two atoms as

$$\text{Kr}\,(^3P_2) + \text{Hg}\,(6\,^1S_0) \qquad\qquad (a)$$
$$s = 1 \qquad\quad s = 0 \qquad\qquad S = 1$$

where s is the resultant spin for each atom and S for the configuration of two atoms. We may suppose the result of the collision to be either

$$\text{Kr}\,(^1S_0) + \text{Hg}\,(8\,^1D_2) \qquad\qquad (b)$$
$$s = 0 \qquad\quad s = 0 \qquad\qquad S = 0$$

or
$$\text{Kr}\,(^1S_0) + \text{Hg}\,(8\,^3D_n) \qquad\qquad (c)$$
$$s = 0 \qquad\quad s = 1 \qquad\qquad S = 1$$

Of the two end states the theorem states that (c) would be the most probable, since the total resultant spin is unchanged.

Beutler and Eisenschimmel[5] investigated spectroscopically the light from a discharge tube containing mercury, krypton and helium. Mercury has two states of nearly the same energy lying close to the $^3P_2$ state of krypton. These are $8\,^3D_n$ and $8\,^1D_2$. They first measured the intensity of the lines emanating from these two states in the mercury-helium discharge containing no krypton. They then added a small amount of krypton to the discharge and measured the intensity of the same lines under these conditions. The results showed that while the intensity of the lines coming from both states increased, that of the line coming from $8\,^3D_n$ increased considerably more than that from $8\,^1D_2$ in accordance with (c). This experiment would seem to substantiate the theory that "electron spin is conserved" on collision.

Sensitized band spectra of several molecules have also been observed. The explanation of the processes involved in the production of such spectra involves a discussion of certain chemical reactions taking place in the excited gases. The production of these bands will therefore be discussed in the next section on the chemical effects of resonance radiation.

3. INTERACTION OF EXCITED ATOMS WITH MOLE-CULES. CHEMICAL REACTIONS TAKING PLACE IN THE PRESENCE OF OPTICALLY EXCITED ATOMS; SENSI-TIZED BAND FLUORESCENCE

3a. INTRODUCTION. It has long been known that certain chemical reactions which will not proceed under the influence of a given frequency of light can be stimulated by adding a substance which is sensitive to this light frequency. After the reaction has run its course, it is found that the added substance is unchanged in composition or physical properties. Such a reaction is said to be "photosensitized" to the frequency ν by the addition of a "sensitizer". As examples of this process we may cite the sensitization of photographic emulsions to green and red light by addition of certain dyestuffs, and also the sensitization of ozone decomposition to certain light frequencies by the addition of chlorine.*

Two problems arise when one studies photosensitized reactions: (1) the action of the sensitizer and (2) the subsequent steps of the chemical reactions occurring as a result of the action of the sensitizer. The first problem has been practically solved through the combined efforts of physicists and chemists, and is intimately connected with the study of resonance radiation. The second problem, however, being of a more complicated nature, is still fraught with difficulties, and many contradictions are to be found in the literature concerning it. We shall be more concerned in this chapter with the primary chemical processes occurring as the result of the absorption of resonance radiation by atoms; the more complicated chemical reactions occurring thereafter will be treated with only so much detail as will serve to give a picture of this field of research.

3b. REACTIONS TAKING PLACE IN THE PRESENCE OF EXCITED MERCURY ATOMS. That atoms excited by resonance radiation may give over their energy to other molecules and cause them to react chemically was first shown by Cario and Franck [13]. They found that hydrogen, activated by collision with excited mercury atoms, could be made to react with

---

* F. Weigert, *Ann. der Phys.* **24**, 243 (1907).

metallic oxides, whereas, under similar temperature conditions and without the presence of activated mercury vapour, no reaction would occur.

The apparatus consisted of a quartz tube containing a drop of mercury and some metallic oxide (CuO or WO_3). The quartz tube was connected to a vacuum system through which pure hydrogen, obtained by heating a palladium tube, could be admitted at low pressures. The pressure of hydrogen could be measured throughout the course of the reaction by means of suitable manometers. Any condensable matter, formed as a result of the reaction, could be frozen out in a liquid-air trap. The temperature of the reaction vessel was maintained at 45° C. during the course of the experiments.

If the reaction tube contained a small amount of hydrogen (a few tenths to 20 mm. pressure) and was illuminated by light from a cooled mercury arc, a decrease in the pressure of hydrogen was observed, whereas, if the mercury arc was allowed to run hot so that the resonance line was reversed, no change in pressure was noted. Furthermore, no reaction was found to occur when mercury vapour was absent. Experiments with the yellow oxide of tungsten (WO_3) showed that it was reduced to the blue oxide in those experiments in which a decrease in the hydrogen pressure occurred. The fact that no reaction occurred unless the incident radiation contained the *un-reversed* 2537 line of mercury showed at once that the first step in the process was the absorption of the resonance line, resulting in the formation of excited ($6\,^3P_1$) mercury atoms (hydrogen shows no absorption in this region of the spectrum). The fact that the oxides were reduced when hydrogen was in the presence of excited mercury atoms, whereas no reaction occurred in their absence, led the authors to suppose that atomic hydrogen was formed as a result of a collision between an excited mercury atom and a hydrogen molecule. This line of reasoning follows from well-known chemical experiments, which show that ordinary molecular hydrogen will not react with CuO or WO_3 at the temperatures employed in the experiment, but that atomic hydrogen, formed in a discharge tube, will reduce these oxides. Cario and Franck suggested, therefore, that the second step of the process was the dissociation of a hydrogen molecule into two atoms as a result of the collision with an excited mercury atom. This process is energetically possible, since the energy of an excited mercury atom is 4·9 volts and that needed to dissociate hydrogen is only 4·46 volts. The remainder of the energy (about 0·4 volt) was supposed to be taken up as relative kinetic energy of the three atoms after collision. The mechanism suggested by Cario and Franck for this step in the process, while energetically possible, is not the only simple mechanism which might be suggested, as we shall show later. It is sufficient for our present purpose simply to state that hydrogen is "activated" by collision of the second kind with excited mercury atoms, thereby being enabled to react with the metallic oxides.

Soon after this experiment was performed other reactions sensitized with mercury vapour and involving the reaction of hydrogen with other chemical elements were studied. Thus Dickinson [18] showed that hydrogen would react with oxygen at low temperatures when a mixture of these gases with mercury vapour was radiated with the unreversed resonance line of mercury. A further study of this reaction was carried out by Mitchell [48], Taylor [67], Marshall [44, 45, 46], and many others. Hydrogen was also found to react with ethylene [44] and with many other substances under the influence of excited mercury vapour. Finally, other substances such as ammonia [19, 69], hydrazine [22] and water [63] were found to decompose in the presence of excited mercury atoms.

It will be of more interest to forgo a chronological discussion of these reactions and establish a few important facts. The fact that none of these sensitized reactions will proceed unless mercury vapour is present and the mixture illuminated with the *unreversed* resonance line of mercury shows that mercury atoms in the $6\,^3P_1$ state are necessary for the process. It does not show, however, that atoms in still higher quantum states, brought there by stepwise excitation, are not involved in the process. That mercury atoms in higher quantum states than the $6\,^3P_1$ are not involved to any appreciable extent in the activating process has been shown by many experiments.

Marshall[44] showed that the introduction of a chlorine-bromine filter, absorbing all radiation between 2900 and 5000 Å., had no effect on the hydrogen-oxygen reaction. Marshall[46] and Frankenburger[25] found that the hydrogen-oxygen and the hydrogen-carbon monoxide reactions are directly proportional to the intensity of the 2537 line in the exciting source. Mitchell and Dickinson[49] observed no increase in the rate of ammonia decomposition when the mixture was radiated with the light from an uncooled mercury arc (emitting radiation longer than 2300) in addition to that from a cooled mercury arc. Finally, Elgin and Taylor[22], by interposing screens of known transmission between the light source and the reaction vessel, showed that the rate of decomposition of hydrazine, in the presence of excited mercury vapour, was proportional to the first power of the light intensity.

That metastable ($6\,^3P_0$) mercury atoms may also activate hydrogen was shown by Meyer[47], who added nitrogen, at about 10 mm. pressure, to a mixture of hydrogen and mercury vapour. He found that the addition of nitrogen increased the rate of reaction between hydrogen and metallic oxides, especially at low hydrogen pressures. The explanation of this effect is analogous to that of the experiments of Donat and Loria on the sensitized fluorescence of thallium. Collisions between nitrogen and $6\,^3P_1$ mercury atoms lead to the formation of $6\,^3P_0$ mercury atoms. The production of the metastable atoms increases the rate of the reaction since, owing to their long life, they have a greater chance of colliding with hydrogen molecules while still excited than have $6\,^3P_1$ atoms, and many of the latter lose their energy by radiation before collision with hydrogen.

Cario and Franck[13] investigated the effect of hydrogen pressure on the rate of the hydrogen-copper oxide reaction. They found that the rate increased as the pressure increased up to a limiting value, above which no increase in rate occurred. The explanation is that at the higher pressures the excited mercury atom loses its activational energy to hydrogen before it has time to radiate. At the lower pressures, however, a certain number of mercury atoms will radiate their energy before colliding, resulting in a falling off in the rate of reaction. A quantitative treatment of this idea was given by Turner [71], which enabled him to obtain a rough estimate of the mean life of a mercury atom in the $6\,^3P_1$ state from the data of Cario and Franck. Since considerations of this type are fundamental to the elucidation of the mechanism of all sensitized reactions, we give the following simple derivation.

Mercury atoms are excited by absorption of radiation from the arc. (1) The number excited is proportional to the intensity of light, I, of wave-length 2537 in the arc and to the concentration of mercury vapour in the tube, [Hg]. Excited mercury atoms either (2) radiate or (3) collide with a hydrogen molecule and activate it. We shall assume that every hydrogen molecule struck eventually leaves the gas phase due to reaction with the oxide. The rate of formation of excited mercury atoms is given by*

$$\frac{d}{dt}[\text{Hg}'] = k_1 I [\text{Hg}] \qquad \ldots\ldots(15),$$

and the rate at which they leave the excited state by

$$-\frac{d}{dt}[\text{Hg}'] = k_2 [\text{Hg}'] + k_3 [\text{Hg}'][\text{H}_2] \qquad \ldots\ldots(16).$$

The rate of activation of hydrogen is given by

$$R = k_3 [\text{Hg}'][\text{H}_2] \qquad \ldots\ldots(17).$$

When a steady state has been reached the rates of formation and destruction of excited mercury atoms will become equal and we may equate (15) and (16), whereby we may solve for the unknown [Hg']. Eq. (17) will then become

$$R = \frac{k_3 k_1 I [\text{Hg}][\text{H}_2]}{k_2 + k_3 [\text{H}_2]} \qquad \ldots\ldots(18),$$

or
$$\frac{1}{R} = \frac{k_2}{k_3 k_1 I [\text{Hg}]} \cdot \frac{1}{[\text{H}_2]} + \frac{1}{k_1 I [\text{Hg}]} \qquad \ldots\ldots(19).$$

From Eq. (19) it will be seen that a plot of $1/R$ against $1/p_{\text{H}_2}$ should give a straight line, the ratio of whose slope to intercept

---

* As is customary in chemical reaction theory we designate the concentration of any substance by [   ]. The partial pressure of any constituent will be proportional to the concentration.

should be k_2/k_3, the constants k_1, I, [Hg] thereby being eliminated. Turner found that a plot of $1/R$ against $1/p_{H_2}$ from Cario and Franck's data did actually give a straight line in agreement with theory. Since k_2 is the chance that a $6\,{}^3P_1$ mercury atom will emit a light quantum, it is obviously equal to $1/\tau$, where τ is the mean life of the excited state. Similarly, k_3 is the chance of collision between an excited mercury atom and a hydrogen molecule, which may be calculated from kinetic theory. The ratio k_2/k_3, determined from this experiment, therefore gives a measure of τ. Turner's calculation gives a value for τ in fair agreement with values obtained by other methods.

It should be emphasized at this point that since the mechanism postulated here does not take account of the imprisonment or diffusion of resonance radiation or the broadening of the absorption line due to pressure, the calculation should only be applied when pressure of the mercury vapour and of the hydrogen is small. Other experiments, which we shall note later, were performed at atmospheric pressure of reacting gases and high mercury vapour pressure, so that such simple considerations do not apply. In order to obtain any real insight into the mechanism of reactions occurring through the agency of resonance radiation, all pressures should be kept as low as possible.

3c. THE MECHANISM OF THE ACTIVATION OF HYDROGEN BY EXCITED MERCURY ATOMS. The question of the mechanism of the activation of hydrogen by excited mercury atoms, which we have so far left open, is one which has interested physicists and chemists for a number of years. Originally three separate mechanisms were proposed: (1) Cario and Franck supposed that a $6\,{}^3P_1$ mercury atom collides with a hydrogen molecule and dissociates it in an elementary act into two hydrogen atoms:

$$\text{Hg}\,(6\,{}^3P_1) + H_2 = \text{Hg}\,(6\,{}^1S_0) + H + H.$$

(2) Mitchell[48] postulated that the result of the collision was a hydrogen molecule in a high state of oscillation and rotation:

$$\text{Hg}\,(6\,{}^3P_1) + H_2 = H_2' + \text{Hg}\,(6\,{}^1S_0).$$

(3) Compton and Turner[17] assumed that the result of the collision was a HgH molecule and a hydrogen atom:

$$Hg\,(6\,^3P_1) + H_2 = (HgH)' + H.$$

All of the above mechanisms are energetically possible.

That some atomic hydrogen is formed as a result of collisions between excited mercury atoms and hydrogen molecules has been shown by Senftleben and his collaborators[62, 64]. They illuminated a quartz vessel containing hydrogen and mercury vapour with light from a water-cooled mercury arc, and arranged to measure the changes in heat conductivity of the hydrogen by a hot-wire method. They found an increase in the heat conductivity when the mixture was illuminated, but on interposing an absorbing layer of mercury vapour between the light source and the conductivity vessel a much smaller increase in the conductivity was observed, showing that the effect was caused by the production of excited mercury atoms. From the increase in heat conductivity they inferred that hydrogen atoms are produced. The experiment is, however, not entirely free from objections, and hence cannot be considered as conclusive proof of the production of hydrogen atoms.

As evidence for the formation of HgH as the mechanism of the collision, Compton and Turner observed that HgH bands were formed in a low voltage arc in hydrogen and mercury. The HgH bands occurred with greatest intensity in those places in the discharge which were shown, by absorption measurements, to contain a large number of $6\,^3P_1$ mercury atoms. More recently Gaviola and Wood[29], investigating sensitized band fluorescence, observed that HgH bands appeared when mixtures of hydrogen, nitrogen and mercury were radiated with light from a mercury arc.

On the theoretical side Compton and Turner have pointed out that the reverse process to (1) would be a very improbable occurrence, since it would involve a simultaneous collision between two hydrogen atoms and one mercury atom. By applying the principle of microscopic reversibility, they concluded that mechanism (1) must likewise be improbable.

Beutler and Rabinowitsch[8] have made a more complete theoretical analysis of the problem. They assume that when an excited atom A' gives over its energy to a molecule BC, not only energy but also linear and angular momentum must be conserved during the process. Before the collision the molecule BC will have a certain amount of rotational energy ($E_1^{rot.}$) and the atom A' will have some kinetic energy relative to the centre of gravity of BC. If the collision takes place in such a way that relative motion of the two particles is not along the line joining the centre of A' with the centre of gravity of BC, A' and BC may be considered to be momentarily rotating about their common centre of gravity, giving rise to an amount of angular momentum M_1. Of course, before the collision the molecule BC has some angular momentum ($M_1^{rot.}$). As a result of the collision the particles must go apart in such a way as to conserve energy and angular momentum. Beutler and Rabinowitsch have shown that if the mass of A is large compared to B or C, the reaction

$$A' + BC = (AB)^{rot.} + C$$

will occur in which $(AB)^{rot.}$ denotes that the molecule AB will possess a large amount of rotational energy. If the reaction is exothermic, they showed that AB would possess more rotational energy than if it were slightly endothermic.

Beutler and Rabinowitsch have found confirmation of these ideas in the experiments of Gaviola and Wood[29]. They found that HgH bands appeared when mercury, nitrogen and a small amount of hydrogen were radiated with the resonance line of mercury. The bands also appeared when mercury and water vapour were similarly illuminated. In the first case lines from the higher rotational states of the HgH molecule were very intense, whereas in the second case (H_2O) the lower rotational states were predominant. In both cases the intensity of the bands was proportional to the square of the incident light intensity, showing that two excited mercury atoms are involved in the process. Since nitrogen or water vapour was present in either case, it may be assumed that metastable mercury atoms were responsible for the fluorescence.

The mechanism of production of the bands, as well as the energy relations, may be written as follows. Taking the heat of dissociation of H_2 as 4·46 volts, of H_2O into OH and H as 5·1 volts, and that of HgH as 0·37 volt, we have

$$Hg\,(6\,^3P_0) + H_2 = HgH + H + 0·62\,\text{volt} \qquad (a),$$
$$Hg\,(6\,^3P_0) + H_2O = HgH + OH - 0·1\,\text{volt} \qquad (b),$$
$$HgH + Hg' = (HgH)' + Hg \qquad (c),$$
$$(HgH)' = h\nu + HgH \qquad (d).$$

That (b) occurs is shown by the presence of OH bands in the fluorescence. Since reaction (a) is exothermic HgH molecules in high rotational states should be produced, if the theory is correct. Since (b) is slightly endothermic HgH molecules in low rotational states should be produced. If it be assumed that the second collision between Hg′ and HgH does not appreciably alter the rotational energy of the molecule, it will be seen that the experimental results are in accord with the theory.

Still more evidence to the effect that mercury hydride and hydrogen are formed as a result of the collision between hydrogen and excited mercury may be drawn from the behaviour of the analogous metals, cadmium and zinc. Bender[4] radiated a mixture of hydrogen and cadmium vapour with light from a hydrogen and cadmium discharge tube emitting the full cadmium spectrum, together with that of hydrogen and cadmium hydride. He noticed that the fluorescent light from the resonance tube contained not only some of the cadmium lines but also CdH bands. Experiment showed that CdH was formed by collision with excited $(5\,^3P_1)$ cadmium atoms, and that the subsequent emission of the CdH bands was due to optical excitation by light from the discharge tube. Furthermore, if the cadmium metal used in the tube contained some oxides, a chemical reaction took place between the atomic hydrogen, formed as a result of the process, and the oxides in the tube. This was shown by the decrease in the hydrogen pressure and by the freezing out of water vapour in a liquid-air trap. If the cadmium metal was distilled free from oxides, the pressure drop was greatly decreased and no water was formed. In this case the formation of two hydrogen atoms as a result of the collision is definitely impossible, since the energy of the $5\,^3P_1$ state of cadmium is only 3·78 volts. The reaction

$$Cd\,(5\,^3P_1) + H_2 = CdH + H$$

is energetically possible, since the heat of dissociation of CdH is 0·67 volt. The sum of the excitation energy of Cd $(5\,^3P_1)$ and the heat of dissociation of CdH $(3·78 + 0·67 = 4·45)$ is practically equal to the dissociation energy of hydrogen (4·46 volts), so that no energy goes into translational energy. A similar reaction takes place when a $4\,^3P_1$ zinc atom collides with a hydrogen molecule forming ZnH and a hydrogen atom. The reaction

$$Zn\,(4\,^3P_1) + H_2 = ZnH + H,$$

as written, goes with the emission of 0·5 volt energy. Bender pointed out that this 0·5 volt may be taken up as vibrational energy of the ZnH molecules.

That the more highly excited $n\,^1P_1$ zinc or cadmium atoms did not enter into the process was shown by the fact that fluorescent lines coming from the $n\,^1P_1$ state were not quenched by the addition of hydrogen. That this state, having an energy considerably in excess of that necessary to dissociate a hydrogen molecule, is not quenched by hydrogen is further evidence supporting the theory that collisions of the second kind are improbable when a large amount of energy goes into translational motion.

Recently, Calvert[10] has succeeded in dissociating hydrogen molecules by impact with optically excited xenon atoms. A reaction vessel, fitted with fluorite windows, containing hydrogen and xenon at low pressures and a small amount of tungsten oxide, WO_3, was irradiated with light from a He-Xe discharge tube. The exciting source gave a very strong xenon spectrum, including the resonance line 1469 $(^1S-^3P_1)$, or in new notation $[(5p^6)\,^1S_0-(5p^5\,.\,6s)\,1^0]$. The xenon atoms in the tube absorbed the resonance line and reached the $^3P_1$ state (energy 8·5 volts), as was shown by the fact that resonance radiation was emitted from the reaction vessel when it contained only xenon. When hydrogen was admitted to the vessel and WO_3

was present, a decrease in the pressure of hydrogen occurred on illumination. The yellow oxide of tungsten was also discoloured, showing that it had been reduced. Since hydrogen itself can absorb no light in the spectral region emitted by the exciting source, it follows that the active form of hydrogen must have been produced by collision with excited xenon atoms. Since there appears to be no possibility of molecule formation, due to the inert nature of xenon, Calvert concludes that the excited xenon atoms must have dissociated the hydrogen molecules, an amount of energy corresponding to 2·6 volts going into relative kinetic energy of the reacting particles. He furthermore supposes that the mechanism of the dissociation of hydrogen by excited mercury atoms is similar to that by excited xenon atoms, in support of the early considerations of Cario and Franck.

3d. REACTIONS INVOLVING HYDROGEN. The reaction between hydrogen and oxygen in the presence of excited mercury atoms has been studied by many investigators. It was at first believed that the primary product of the reaction was water. Later studies by Marshall[44, 45, 46], Hirst and Rideal[31, 32], Bonhoeffer and Loeb[9], Frankenburger and Klinkhardt[26, 36] showed that hydrogen peroxide (H_2O_2) was formed in great quantities during the reaction. The greatest yields of H_2O_2 were obtained when the experiment was performed by flowing the gases through the illuminated reaction zone. Taylor postulated a chain mechanism for the reaction between H_2 and O_2. Marshall, by measuring the number of molecules of H_2O_2 formed per quantum of 2537 absorbed, obtained a yield of about four molecules per quantum of resonance radiation absorbed. This result was believed to confirm the chain mechanism of the reaction. Later measurements of the efficiency of this reaction by Frankenburger and Klinkhardt gave a yield of about 1·2 molecules of H_2O_2 per quantum absorbed at 60° C. These authors believe their experiments to show that no chain mechanism is involved. Recent experiments of Bates and Salley[1] appear to confirm the earlier results of Taylor and Marshall.

The sensitized reaction between hydrogen and ethylene (C_2H_4) appears to be quite complicated. Apart from the polymerization of ethylene into solid products, ethane (C_2H_6), acetylene (C_2H_2) and many other hydrocarbons have been found as products of the reaction [2, 15, 58, 67]. Taylor and Hill [70], in a critical presentation of their data and those of other experimenters, appear to believe that the reaction is too complicated to be amenable to theoretical treatment.

The reaction between H_2 and CO, in the presence of excited mercury vapour, is somewhat simpler. Taylor and Marshall found formaldehyde (HCHO) as a product. More recently Frankenburger [25] has found, by taking absorption spectra of the products of the reaction, that both formaldehyde and glyoxal (CHO.CHO) are formed. No unstable intermediate products were found in the spectroscopic analysis. Both Marshall and Frankenburger have measured the photochemical efficiency of the reaction. The former found a yield of more than five molecules of HCHO per quantum absorbed, while the latter, using an improved form of apparatus, found a yield of about one molecule per quantum. Both observers agree, however, that H_2 and CO disappear in about a 1 : 1 ratio.

Taylor and Marshall have shown that the reaction between H_2 and N_2O proceeds rapidly in the presence of excited mercury atoms, whereas that between H_2 and CO_2 does not. Most observers agree that no reaction between H_2 and N_2 occurs under the action of excited mercury atoms. Hirst and Rideal, and Noyes [55], on the other hand, have found a reaction to occur, but their experiments were not carried out under circumstances from which one could conclude with certainty that excited mercury atoms were the activating agent.

3e. THE SENSITIZED FORMATION OF OZONE. Of those sensitized reactions not involving hydrogen the first to be studied was the formation of ozone. Dickinson and Sherrill [20] allowed oxygen at atmospheric pressure to become saturated with mercury vapour at 20° C. and to flow through a chamber illuminated with the light of a water-cooled quartz mercury arc. To eliminate the direct photochemical formation of ozone, which occurs under the influence of radiation in the region below 2000 Å., a tartaric acid filter was used to absorb this radiation from the arc. Their results showed that ozone was formed in the presence of excited mercury atoms. The reaction is also complicated by the formation of HgO. They showed, however, that the reaction

$$Hg' + O_2 = HgO + O$$

could not be the first step in the process, since at least seven molecules of ozone were formed per mercury atom used. Their mechanism assumes that an excited oxygen molecule is formed as a result of the collision between excited mercury and normal oxygen, and that the formation of ozone proceeds by the following mechanism:

$$Hg' + O_2 = O_2' + Hg,$$
$$O_2' + O_2 = O_3 + O,$$
$$O + O_2 = O_3.$$

It is possible, however, that the first step in the reaction is

$$Hg' + O_2 = Hg + O + O,$$

since the latest value for the heat of dissociation of the oxygen molecule into two normal atoms is 5·09 volts. The excess energy required for dissociation above that supplied by the excited mercury atom might be derived from the relative kinetic energy of the mercury atom and oxygen molecule. The formation of HgO is probably the result of the action of ozone on normal mercury atoms. Leipunsky and Sagulin [42] believed that HgO was formed as a primary product. They used no light filter, however, and certainly had some photochemical formation of ozone as a result of short-wave-length ultra-violet light. The work of Noyes [56], performed at low pressures and with the help of filters, would appear to substantiate that of Dickinson and Sherrill in most respects.

3f. THE SENSITIZED DECOMPOSITION OF AMMONIA. Another very interesting reaction is the mercury-sensitized decomposition of ammonia. This reaction has been studied extensively by Dickinson and Mitchell [49, 50], and by Bates and Taylor [2, 69]. Dickinson and Mitchell studied the reaction at low pressures (0·1 to 3 mm.) and used an acetic acid filter to eliminate the direct photochemical decomposition of ammonia which takes place under the influence of radiation of wavelength shorter than 2300. They showed by the usual methods that decomposition occurs in the presence of excited mercury atoms. Analysis of the decomposition products was made, after freezing out undecomposed ammonia, by the use of a quartz-fibre manometer and McLeod gauge. The combined use of such apparatus enables one to measure the average molecular weight of the decomposition products at low pressures. The result of the analysis showed 70 per cent. hydrogen and 30 per cent. nitrogen, or practically a stoichiometric ratio (75:25).*

Mitchell and Dickinson investigated further the effect of nitrogen, argon and hydrogen on the rate of decomposition. Argon and nitrogen at pressures up to 0·4 mm. had no effect on the rate, as would be expected from the fact that at these pressures they do not quench mercury resonance radiation appreciably. Hydrogen, on the other hand, had a decided quenching effect on the rate of decomposition, even a greater effect than one would be led to expect from its known quenching ability. The authors inferred from this that excited ammonia molecules must be involved in the reaction, and furthermore, that hydrogen may take excitation energy from them as well as from excited mercury atoms. An analysis of their experimental data by methods similar to those already described in this chapter appears to substantiate this view.

Evidence for the existence of excited ammonia molecules is given by the discovery of Dickinson and Mitchell that mixtures of ammonia and mercury exhibited a fluorescence when radiated with the light of a cooled mercury arc. Mitchell[50] showed that the fluorescence consisted of a diffuse band stretching from about 2600 up into the green, with a maximum at about 3400 Å. Gaviola and Wood further investigated the fluorescence and showed that the intensity of the band was propor-

* Bates and Taylor, using a flow system and no light filter, found a larger percentage of hydrogen in their products, indicating the formation of hydrazine and hydrogen. More recent work of Elgin and Taylor[22] (see Table III, p. 2069) would appear to substantiate the results of Dickinson and Mitchell.

tional to the first power of the intensity of the exciting light, demonstrating that only one excited mercury atom is involved in its production. Mitchell further showed that the intensity of the maximum of the band increased with increasing ammonia pressure in an analogous way to which the rate of decomposition increased with ammonia pressure. Similarly, at constant ammonia pressure, the intensity of the maximum of the band decreased with hydrogen pressure in the same way as did the rate of decomposition under like circumstances. These results show that the decomposition is intimately connected with the fluorescence.

3g. OTHER DECOMPOSITIONS SENSITIZED BY EXCITED MERCURY ATOMS. The decomposition of a great number of substances has been investigated by Bates and Taylor[2, 69], a compendium of whose results is shown in Table VIII. Of all

TABLE VIII

Substance	Ratio of photosensitized to non-photosensitized rate of reaction	Analysis of gaseous products of the non-photosensitized reaction	Analysis of gaseous products of the photosensitized reaction
H_2O	—	—	73 % H_2; 27 % O_2
NH_3	200 : 1	96 % H_2; 4 % N_2	89 % H_2; 11 % N_2
C_2H_4	—	—	88 % H_2; 12 % CH_4, etc.
CH_3OH	600 : 1	—	58 % H_2; 42 % $CH_4 + CO$
C_2H_5OH	50 : 1	—	46 % H_2; 50 % $CO + CH_4$
C_6H_{14}	1000 : 1	—	96 % H_2; 4 % CH_4
C_6H_6	30 : 1	—	60 % H_2; 40 % CH_4
$(CH_3)_2CO$	2 : 1	100 % $CO + CH_4$	100 % $CO + CH_4$
$HCOOH$	400 : 1	—	76 % CO; 24 % H_2
$C_2H_5NH_2$	60 : 1	96 % H_2; 4 % N_2	96 % H_2; 3·7 % CH_4; 0·3 % N_2

the substances investigated, none failed to decompose in the presence of excited mercury atoms. The analysis of the gaseous decomposition products is shown in column 4 of the table. The decomposition of water was investigated earlier by Senftleben and Rehren[63], who found, however, no oxygen in their

decomposition products. It is quite probable that OH is a
product of the elementary process according to the mechanism

$$Hg' + H_2O = Hg + H + OH$$
$$\text{or} \qquad\qquad = HgH + OH,$$

since OH bands have been found in sensitized fluorescence in
the presence of mercury and water vapour.

Finally, Elgin and Taylor have investigated the photosensitized decomposition of hydrazine (N_2H_4) and found N_2,
H_2 and NH_3 as decomposition products. They also showed that
the addition of 200 mm. of hydrogen (initial pressure of hydrazine 10 mm.) made no change in the rate of the reaction. Since
this effect is not to be expected on account of the known
quenching power of hydrogen for mercury resonance radiation,
they were able to give no explanation of it. Nitrogen and
ammonia also showed no marked effect on the rate. It would
be of interest to investigate this reaction at low pressures
where broadening of the absorption line does not occur.

3 h. REACTIONS SENSITIZED BY OTHER METALLIC VAPOURS
ACTIVATED BY THE ABSORPTION OF RESONANCE RADIATION.
We have already mentioned that Bender showed that the
reaction between hydrogen and metallic oxides occurs in the
presence of excited zinc or cadmium atoms. Bates and
Taylor[3] found no reaction to occur between hydrogen and
ethylene in the presence of excited cadmium atoms. He also
could observe no ammonia decomposition under like circumstances. The investigation of reactions sensitized by cadmium
and zinc resonance radiation is difficult, since it was not easy to
get a very intense source of the resonance radiation of these
metals, and since, furthermore, at the temperatures necessary
to obtain a sufficient vapour pressure of these elements many
reactions proceed thermally.

As further evidence of the fact that a collision of the second
kind is most effective where least energy goes into kinetic
energy, Beutler and Eisenschimmel[6] showed that excited
neon atoms (from a discharge tube) would dissociate hydrogen
into a normal and an excited atom according to the scheme

$$Ne\,(^3P_2) + H_2 = Ne\,(^1S_0) + H + H\,(3\,^2P) + 0.13\,\text{volt}.$$

As evidence of this reaction they showed that the H_α line of hydrogen ($2\,^2S$–$3\,^2P$) appeared with high intensity in the discharge tube containing neon and hydrogen.

4. BANDS CONNECTED WITH RESONANCE LINES

There are two types of band fluorescence which appear to be intimately connected with processes giving rise to resonance radiation: (a) bands which lie close to the resonance line and appear in fluorescence when a mixture of mercury vapour and certain rare gases is illuminated by the unreversed 2537 line; and (b) diffuse bands, or continua stretching throughout a large region of the spectrum, which occur when mercury vapour at high temperature and pressure is optically excited. In addition to these two classes, there are bands which may be definitely ascribed to the molecules Hg_2, Cd_2, Zn_2, a description of which may be found in the literature on band spectra.

4a. MERCURY-RARE GAS BANDS. Oldenberg[57] made an investigation on mixtures of mercury vapour with the rare gases helium, neon, argon, krypton and xenon. A mixture of mercury with one of the rare gases (at high pressure) was illuminated by a cooled mercury arc. The fluorescence consisted of the resonance line, together with certain continua and bands lying in the neighbourhood of 2537. That the fluorescence is connected with the production of excited mercury atoms was shown by the fact that no fluorescence occurred when the core of the resonance line was absent (exciting lamp operated without cooling).

Oldenberg's experiments were made to test the following idea: Suppose an excited mercury atom collides with a foreign gas atom and radiates at the moment of collision. Can the excitation energy of the mercury atom co-operate in such a way with the relative translational energy of the two atoms that a line of wave-length different from 2537 will be radiated? The process would correspond to the following equation:

Energy of excited mercury $\pm$ Translational energy
$$= h\nu \text{ (radiated)}.$$

Since the translational energy is, in general, not definite, one would expect to find a continuous band of radiation extending a short distance to either side of the 2537 line. Oldenberg actually found that such continua did occur when mixtures of $Hg + Ne$, $Hg + He$ and $Hg + Xe$ were examined. The maximum extension of the continuum toward the short wave-length side from the 2537 line was found to be greatest for the lightest gases, in agreement with predictions made from kinetic theory.

In addition to the continuous spectra observed, Oldenberg found discrete bands lying to the long wave-length side of 2537, when mixtures of $Hg + A$ or $Hg + Kr$ were optically excited. There were from five to seven bands observed and their position was found to depend on whether argon or krypton was present. From an intensive study of the subject, the author came to the conclusion that these bands are due to the unstable molecules HgA and HgKr respectively. Since it is known that mercury and the rare gases do not combine to a measurable extent when both are unexcited, Oldenberg supposed that an excited mercury atom might form a loosely bound molecule, due to polarization forces, with an argon or krypton atom. Since the binding energy of the combinations is small, the light radiated will have a frequency quite near to that of the mercury resonance line.

In some cases, notably argon, krypton and xenon, Oldenberg found that the continuum to the short wave-length side of 2537 consisted of two broad diffuse maxima, the distance between the maxima depending on the rare gas used. The explanation of this effect, due to Kuhn and Oldenberg[40], is that collision between an excited mercury atom and a normal rare gas atom results in the formation of quasi-molecules which may be in either of two vibrational states, depending on the direction of the relative momenta at the time of collision. From each of these vibrational states a band may be radiated when the molecule returns to the normal state, thus explaining the two maxima observed. The fact that two maxima occur is evidence in favour of space quantization upon impact.

4b. CONTINUA APPARENTLY ASSOCIATED WITH RESONANCE RADIATION. In addition to the bands discussed in the foregoing section, continua, or structureless diffuse bands, have been observed when the vapours of mercury, cadmium and zinc are optically excited. Since the fluorescence shows no structure, it has been exceedingly difficult to find a mechanism which will explain its production. A large number of experiments have been performed for the purpose of explaining the phenomena involved, but none appears to have given a completely satisfactory explanation. We shall content ourselves here, therefore, with giving a short description of the phenomena.

If mercury vapour, at high pressure, be excited by a high frequency discharge, or by light of a wave-length near to the resonance line, three sets of continua appear, viz. (1) a continuum, extending a short distance to the long wave-length side of 2537; (2) a broad continuum having a maximum at 3300; and finally, (3) a continuum with a maximum at about 4850. Experiments of van der Lingen and Wood [72] indicate that the bands only appear in distilling vapour, but other observers do not substantiate this result. If the bands are optically excited by the 2537 line, their intensity is proportional to the first power of the intensity of the exciting light [60]. The general belief is that the bands are due to loosely bound mercury molecules. In support of this theory, Houtermans [33] obtained evidence showing that the bands are connected with mercury atoms in the $6\,^3P_1$ and $6\,^3P_0$ states.

Similar diffuse bands appear in the spectra of the elements cadmium, zinc, thallium and magnesium. A compendium of the results of various researches on the subject is given by Mrozowski [52] and by Hamada [30]. As in the case of mercury, the bands appear to be due to loosely bound molecules and are associated with the various excited states of the atoms forming the molecule.

REFERENCES TO CHAPTER II

[1] Bates, J. R. and Salley, D. J., *Journ. Amer. Chem. Soc.* **55**, 110 (1933).
[2] Bates, J. R. and Taylor, H. S., *ibid.* **49**, 2438 (1927).
[3] —— —— *ibid.* **50**, 771 (1928).
[4] Bender, P., *Phys. Rev.* **36**, 1535, 1543 (1930).

[5] Beutler, H. and Eisenschimmel, W., *Z. f. Phys. Chem.* B **10**, 89 (1930).
[6] —— —— *Z. f. Elektrochem.* **37**, 582 (1931).
[7] Beutler, H. and Josephy, B., *Z. f. Phys.* **53**, 747 (1929).
[8] Beutler, H. and Rabinowitsch, E., *Z. f. Phys. Chem.* B **8**, 231, 403 (1930).
[9] Bonhoeffer, K. F. and Loeb, S., *ibid.* **119**, 474 (1926).
[10] Calvert, H. R., *Z. f. Phys.* **78**, 479 (1932).
[11] Carelli, A., *ibid.* **53**, 210 (1929).
[12] Cario, G., *ibid.* **10**, 185 (1922).
[13] Cario, G. and Franck, J., *ibid.* **11**, 161 (1922).
[14] —— —— *ibid.* **17**, 202 (1923).
[15] Clemenc, A. and Patat, F., *Z. f. Phys. Chem.* B **3**, 289 (1929).
[16] Collins, E. H., *Phys. Rev.* **32**, 753 (1928).
[17] Compton, K. T. and Turner, L. A., *Phil. Mag.* **48**, 360 (1924).
[18] Dickinson, R. G., *Proc. Nat. Acad. Sci.* **10**, 409 (1924).
[19] Dickinson, R. G. and Mitchell, A. C. G., *ibid.* **12**, 692 (1926).
[20] Dickinson, R. G. and Sherrill, M. S., *ibid.* **12**, 175 (1926).
[21] Donat, K., *Z. f. Phys.* **29**, 345 (1924).
[22] Elgin, J. C. and Taylor, H. S., *Journ. Amer. Chem. Soc.* **51**, 2059 (1929).
[23] Franck, J. and Jordan, P., *Anregung von Quantensprüngen durch Stösse*, J. Springer, Berlin, pp. 226–228.
[24] Franck, J., *Z. f. Phys.* **9**, 259 (1922).
[25] Frankenburger, W., *Z. f. Elektrochem.* **36**, 757 (1930).
[26] Frankenburger, W. and Klinkhardt, H., *Z. f. Phys. Chem.* B **15**, 421 (1931).
[27] Füchtbauer, C., *Phys. Z.* **21**, 635 (1920).
[28] Gaviola, E., *Phil. Mag.* **6**, 1154, 1167 (1928).
[29] Gaviola, E. and Wood, R. W., *ibid.* **6**, 1191 (1928).
[30] Hamada, H., *Nature*, **127**, 554 (1931).
[31] Hirst, H. S. and Rideal, E. K., *ibid.* **116**, 899 (1925); **117**, 449 (1926).
[32] Hirst, H. S., *Proc. Camb. Phil. Soc.* **23**, 162 (1926).
[33] Houtermans, F. G., *Z. f. Phys.* **41**, 140 (1927).
[34] Kallmann, H. and London, F., *Z. f. Phys. Chem.* B **2**, 207 (1929).
[35] Klein, O. and Rosseland, S., *Z. f. Phys.* **4**, 46 (1921).
[36] Klinkhardt, H. and Frankenburger, W., *Z. f. Phys. Chem.* B **8**, 138 (1930).
[37] Klumb, H. and Pringsheim, P., *Z. f. Phys.* **52**, 610 (1928).
[38] Kopfermann, H., *ibid.* **21**, 316 (1924).
[39] Kopfermann, H. and Ladenburg, R., *ibid.* **48**, 15, 26, 51, 192 (1928); **65**, 167 (1930).
[40] Kuhn, H. and Oldenberg, O., *Phys. Rev.* **41**, 72 (1932).
[41] Latyscheff, G. D. and Leipunsky, A. I., *Z. f. Phys.* **65**, 111 (1930).
[42] Leipunsky, A. and Sagulin, A., *Z. f. Phys. Chem.* B **1**, 362 (1928).
[43] Loria, S., *Phys. Rev.* **26**, 573 (1925).
[44] Marshall, A. L., *Journ. Phys. Chem.* **30**, 34 (1926).
[45] —— *ibid.* **30**, 1078 (1926).
[46] —— *Journ. Amer. Chem. Soc.* **49**, 2763 (1927).
[47] Meyer, E., *Z. f. Phys.* **37**, 639 (1926).

[48] Mitchell, A. C. G., *Proc. Nat. Acad. Sci.* **11**, 458 (1925).
[49] Mitchell, A. C. G. and Dickinson, R. G., *Journ. Amer. Chem. Soc.* **49**, 1478 (1927).
[50] Mitchell, A. C. G., *ibid.* **49**, 2699 (1927); *Journ. Frankl. Inst.* **212**, 341 (1931).
[51] Mohler, F. L., *Bur. Stand. Journ. Res.* **9**, 493 (1932).
[52] Mrozowski, S., *Z. f. Phys.* **62**, 314 (1930).
[53] —— *ibid.* **78**, 826 (1932).
[54] Nordheim, L., *ibid.* **36**, 496 (1926).
[55] Noyes, W. A., Jr., *Journ. Amer. Chem. Soc.* **47**, 1003 (1925).
[56] —— *ibid.* **49**, 3100 (1927); *Z. f. Phys. Chem.* B **2**, 445 (1929).
[57] Oldenberg, O., *Z. f. Phys.* **47**, 184 (1928); **51**, 605 (1928); **55**, 1 (1929).
[58] Olson, A. R. and Meyers, C. H., *Journ. Amer. Chem. Soc.* **48**, 389 (1926); **49**, 3131 (1927).
[59] Orthmann, W. and Pringsheim, P., *Z. f. Phys.* **35**, 626 (1926).
[60] Pringsheim, P. and Terenin, A., *ibid.* **47**, 330 (1928).
[61] Rasetti, F., *Nature*, **118**, 47 (1926).
[62] Senftleben, H., *Z. f. Phys.* **32**, 922 (1925); **33**, 871 (1925).
[63] Senftleben, H. and Rehren, I., *ibid.* **37**, 529 (1926).
[64] Senftleben, H. and Riechemeier, O., *Ann. d. Phys.* **6**, 105 (1930).
[65] Smyth, H. D., *Proc. Nat. Acad. Sci.* **11**, 679 (1925); *Phys. Rev.* **27**, 108 (1926).
[66] Terenin, A., *Z. f. Phys.* **37**, 98 (1926).
[67] Taylor, H. S. and Marshall, A. L., *Journ. Phys. Chem.* **29**, 1140 (1925).
[68] Taylor, H. S., *Trans. Farad. Soc.* **21**, 560 (1925).
[69] Taylor, H. S. and Bates, J. R., *Proc. Nat. Acad. Sci.* **12**, 714 (1926).
[70] Taylor, H. S. and Hill, D. G., *Journ. Amer. Chem. Soc.* **51**, 2922 (1929).
[71] Turner, L. A., *Phys. Rev.* **23**, 466 (1924).
[72] Van der Lingen, J. S. and Wood, R. W., *Astrophys. J.* **54**, 149 (1921).
[73] Webb, H. W. and Wang, S. C., *Phys. Rev.* **33**, 329 (1929).
[74] Winans, J. G., *ibid.* **30**, 1 (1927).
[75] Wood, R. W., *Proc. Roy. Soc.* **106**, 679 (1924); *Phil. Mag.* **50**, 775 (1925); **4**, 466 (1927).
[76] Wood, R. W. and Gaviola, E., *Phil. Mag.* **6**, 271, 352 (1928).

CHAPTER III

ABSORPTION LINES AND MEASUREMENTS OF THE LIFETIME OF THE RESONANCE STATE

1. GENERAL PROPERTIES OF ABSORPTION LINES

Chapters I and II contained, for the most part, descriptions and interpretations of qualitative experiments on resonance radiation and resonance lines. It is the purpose of this chapter to introduce and perfect the physical and mathematical tools which allow a quantitative interpretation of another group of experiments on resonance radiation which, either directly or indirectly, are connected with the formation of absorption lines.

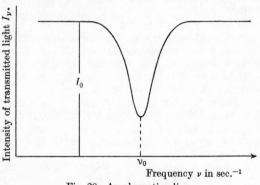

Fig. 20. An absorption line.

1 *a*. THE NOTION OF AN ABSORPTION LINE. If parallel light from a source emitting a continuous spectrum be sent through an absorption cell containing a monatomic gas, the intensity of the transmitted light, I_ν, may show a frequency distribution similar to that depicted in Fig. 20. When this is the case, the gas is said to possess an absorption line at the frequency ν_0, where ν_0 is the frequency at the centre of the line in sec.$^{-1}$. The absorption coefficient k_ν of the gas is defined by the equation

$$I_\nu = I_0 e^{-k_\nu x} \qquad \dots\dots(20),$$

where x is the thickness of the absorbing layer. When x is measured in cm., k_ν is expressed in cm.$^{-1}$. From Fig. 20 and Eq. (20) we may obtain k_ν as a function of frequency, and when this is done we have a curve such as that shown in Fig. 21. The *total breadth* of this curve at the place where k_ν has fallen to one-half of its maximum value, $k_{max.}$, is called the *half breadth of the absorption line* and is denoted by $\Delta\nu$. In general the absorption coefficient of a gas is given by an expression involving a function of ν and a definite value of $k_{max.}$ and $\Delta\nu$, all of which may depend on the nature of the molecules of the gas, their

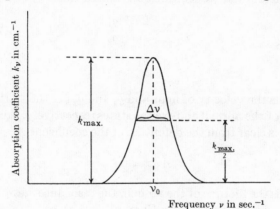

Fig. 21. Variation of absorption coefficient with frequency
in an absorption line.

motion, and their interaction either with one another or with foreign molecules.

1*b*. THE EINSTEIN THEORY OF RADIATION. Consider an enclosure containing isotropic radiation of frequency between ν and $\nu + d\nu$, intensity I_ν and atoms capable of being raised by absorption of the radiation from the normal state 1 to the excited state 2. Following Milne's [47] treatment of the Einstein theory of radiation, we define the following probability coefficients:

$B_{1\to2}I_\nu =$ probability per second that the atom in state 1, exposed to isotropic radiation of frequency between ν and $\nu + d\nu$ and intensity I_ν, will absorb a quantum $h\nu$ and pass to the state 2.

$A_{2\to1}$ = probability per second that the atom in the state 2 will spontaneously emit, in a random direction, a quantum $h\nu$ and pass to the state 1.

$B_{2\to1} I_\nu$ = probability per second that the atom will undergo the transition from 2 to 1 when it is exposed to isotropic radiation of frequency between ν and $\nu + d\nu$ and intensity I_ν, emitting thereby a quantum in the same direction as the stimulating quantum.

By considering the thermodynamic equilibrium between the radiation and the atoms, Einstein showed that

$$\frac{A_{2\to1}}{B_{1\to2}} = \frac{2h\nu^3}{c^2} \cdot \frac{g_1}{g_2} \qquad \ldots\ldots(21),$$

and

$$\frac{B_{2\to1}}{B_{1\to2}} = \frac{g_1}{g_2} \qquad \ldots\ldots(22),$$

where c is the velocity of light and g_1 and g_2 are the statistical weights of the normal and excited states respectively. Furthermore, it is clear from the definition of the coefficient $A_{2\to1}$ that

$$A_{2\to1} = \frac{1}{\tau} \qquad \ldots\ldots(23),$$

where τ is the lifetime of the atom in the resonance state in the sense in which it was used in Chap. I.

It should be emphasized at this point that the B coefficients have been defined in terms of intensity of isotropic radiation, whereas the original Einstein B coefficients were defined in terms of radiation density. The relation between the two kinds of B's is

$$B\,(\text{density}) = \frac{c}{4\pi}\, B\,(\text{intensity}).$$

The Einstein theory of radiation lends itself very naturally to calculations concerning the absorption of light by atoms and molecules, that is, to calculations involving absorption lines. In 1920 Füchtbauer[10] derived a relation between the integral of the absorption coefficient of a line (the area under the curve in Fig. 21) and a probability coefficient connected with the Einstein A coefficient. In 1921 Ladenburg[37] gave a more precise relation between this integral and the Einstein A

coefficient. In 1924 Tolman[74] and Milne[47] derived the relation independently. In the following paragraph a derivation of the formula is given which follows the notation of Milne.

Consider a parallel beam of light of frequency between ν and $\nu + d\nu$ and intensity I_ν travelling, in the positive x direction, through a layer of atoms bounded by the planes at x and $x + dx$. Suppose there are N normal atoms per c.c. of which δN_ν are capable of absorbing the frequency range between ν and $\nu + d\nu$, and N' excited atoms of which $\delta N_\nu'$ are capable of emitting this frequency range. Neglecting the effect of spontaneous re-emission in view of the fact that it takes place in all directions, the decrease in energy of the beam is given by

$$-d[I_\nu \delta\nu] = \delta N_\nu dx h\nu B_{1\to2}\frac{I_\nu}{4\pi} - \delta N_\nu' dx h\nu B_{2\to1}\frac{I_\nu}{4\pi}...(24),$$

where $I_\nu/4\pi$ is the intensity of the equivalent isotropic radiation for which $B_{1\to2}$ and $B_{2\to1}$ are defined. Rewriting Eq. (24), we obtain

$$-\frac{1}{I_\nu}\frac{dI_\nu}{dx}\delta\nu = \frac{h\nu}{4\pi}(B_{1\to2}\delta N_\nu - B_{2\to1}\delta N_\nu')......(25).$$

Recognizing that the left-hand member is $k_\nu \delta\nu$ as defined by Eq. (20), Eq. (25) becomes

$$k_\nu \delta\nu = \frac{h\nu}{4\pi}(B_{1\to2}\delta N_\nu - B_{2\to1}\delta N_\nu'),$$

and integrating over the whole absorption line, neglecting the slight variation in ν throughout the line,

$$\int k_\nu d\nu = \frac{h\nu_0}{4\pi}(B_{1\to2}N - B_{2\to1}N') \quad ......(26),$$

where ν_0 is the frequency at the centre of the line. Making use of Eqs. (21), (22) and (23), we have finally

$$\int k_\nu d\nu = \frac{\lambda_0^2 g_2}{8\pi g_1}\cdot\frac{N}{\tau}\left(1 - \frac{g_1}{g_2}\cdot\frac{N'}{N}\right) \quad ......(27).$$

In gases electrically excited at high current densities, the number of excited atoms may become an appreciable fraction of the number of normal atoms, in which case the quantity

$\dfrac{g_1}{g_2}\cdot\dfrac{N'}{N}$ cannot be neglected. If, however, the only agency responsible for the formation of excited atoms is the absorption of the beam of light itself, the ratio N'/N is exceedingly small, of the order of 10^{-4} or less, and consequently Eq. (27) may be written

$$\int k_\nu\,d\nu=\frac{\lambda_0{}^2 g_2}{8\pi g_1}\cdot\frac{N}{\tau}\qquad\ldots\ldots(28).$$

Eq. (28) is of fundamental importance. It expresses the fact that *whatever physical processes are responsible for the formation of the absorption line, the integral of the absorption coefficient remains constant when N is constant.*

1c. THE RELATION BETWEEN f-VALUE AND LIFETIME. On the basis of the classical electron theory of dispersion, the optical behaviour of N atoms per c.c. was represented by the behaviour of $\mathfrak{N}$ quasi-elastically bound electrons (the so-called "dispersion electrons"). The ratio $\mathfrak{N}/N$ was found to be constant for a particular spectral line and was denoted by f. The f-value associated with a spectral line emitted by an atom can be regarded as a measure of the degree to which the ability of the atom to absorb and emit this line resembles such an ability on the part of a classical oscillating electron. In all classical formulas of normal and anomalous dispersion, magneto-rotation and absorption, the quantity f appears. On the basis of the quantum theory, the f-value has a very simple interpretation: it is proportional to the Einstein A coefficient, or, in the case of a resonance line, it is inversely proportional to the lifetime of the resonance level. This is most easily shown by the classical formula, developed long before the Einstein theory, namely

$$\int\frac{4\pi}{\lambda_0}\,(n\kappa)\,d\nu=\frac{\pi e^2}{mc}\,\mathfrak{N}=\frac{\pi e^2}{mc}\,Nf\qquad\ldots\ldots(29),$$

in which n is the index of refraction and $n\kappa$ the electron theory absorption coefficient which is connected with the usual absorption coefficient by the relation

$$\frac{4\pi}{\lambda_0}\,(n\kappa)=k_\nu\qquad\ldots\ldots(30).$$

From Eqs. (28) and (29) we have Ladenburg's formula[37] stating the connection between the f-value of a resonance line and the lifetime of the resonance level:

$$\frac{\pi e^2}{mc} \cdot Nf = \frac{\lambda_0^2 g_2}{8\pi g_1} \cdot \frac{N}{\tau}$$

or

$$f\tau = \frac{mc}{8\pi^2 e^2} \cdot \frac{g_2}{g_1} \lambda_0^2 = 1 \cdot 51 \frac{g_2}{g_1} \lambda_0^2 \qquad \ldots\ldots(31).$$

Eq. (31) enables us to calculate the lifetime once the f-value has been measured, or vice versa. In Table IX are given values of $f\tau$ for those resonance lines that are most often studied to-day.

TABLE IX

Element	Resonance line	g_2/g_1	λ_0 in Å. units	$f\tau = 1\cdot51\ (g_2/g_1)\ \lambda_0^2 \times 10^9$
Li	$2\,^2S_{1/2}-2\,^2P_{1/2}$	1	6708	6·80
,,	$2\,^2S_{1/2}-2\,^2P_{3/2}$	2	6708	13·6
Na	$3\,^2S_{1/2}-3\,^2P_{1/2}$	1	5896	5·24
,,	$3\,^2S_{1/2}-3\,^2P_{3/2}$	2	5890	10·46
K	$4\,^2S_{1/2}-4\,^2P_{1/2}$	1	7699	8·94
,,	$4\,^2S_{1/2}-4\,^2P_{3/2}$	2	7665	17·8
Cs	$6\,^2S_{1/2}-6\,^2P_{1/2}$	1	8944	12·1
,,	$6\,^2S_{1/2}-6\,^2P_{3/2}$	2	8521	21·9
Mg	$3\,^1S_0\ -3\,^3P_1$	3	4571	9·48
,,	$3\,^1S_0\ -3\,^1P_1$	3	2852	3·68
Ca	$4\,^1S_0\ -4\,^3P_1$	3	6573	19·5
,,	$4\,^1S_0\ -4\,^1P_1$	3	4227	8·09
Zn	$4\,^1S_0\ -4\,^3P_1$	3	3076	4·28
,,	$4\,^1S_0\ -4\,^1P_1$	3	2139	2·07
Sr	$5\,^1S_0\ -5\,^3P_1$	3	6893	21·5
,,	$5\,^1S_0\ -5\,^1P_1$	3	4608	9·64
Cd	$5\,^1S_0\ -5\,^3P_1$	3	3261	4·80
,,	$5\,^1S_0\ -5\,^1P_1$	3	2288	2·37
Ba	$6\,^1S_0\ -6\,^3P_1$	3	7911	28·3
,,	$6\,^1S_0\ -6\,^1P_1$	3	5536	13·9
Hg	$6\,^1S_0\ -6\,^3P_1$	3	2537	2·91
,,	$6\,^1S_0\ -6\,^1P_1$	3	1850	1·55
Tl	$6\,^2P_{1/2}-7\,^2S_{1/2}$	1	3776	$f/A = 2\cdot15$
,,	$6\,^2P_{1/2}-6\,^2D_{3/2}$	2	2768	$f/A = 2\cdot31$
,,	$6\,^2P_{3/2}-7\,^2S_{1/2}$	$\frac{1}{2}$	5350	$f/A = 2\cdot20$

2. THE ABSORPTION COEFFICIENT OF A GAS

2a. EXPRESSION FOR THE ABSORPTION COEFFICIENT. There are in general five processes that contribute to the formation of an absorption line of a gas. Each process can be regarded as an agent for broadening the absorption line. The five types of broadening are as follows:

(1) *Natural* broadening due to the finite lifetime of the excited state.

(2) *Doppler effect* broadening due to the motions of the atoms.

(3) *Lorentz* broadening due to collisions with foreign gases.

(4) *Holtsmark* broadening due to collisions with other absorbing atoms of the same kind.

(5) *Stark effect* broadening due to collisions with electrons and ions.

Both Lorentz and Holtsmark types of broadening are often referred to as "pressure-broadening" ("Druckverbreiterung"), since the first depends on the pressure of the foreign gas, and the second on the pressure of the absorbing gas. Although the recent work of Weisskopf[83] seems to indicate that the two kinds of broadening are identical, this point is still in sufficient doubt to make it desirable to retain the old nomenclature and to distinguish between the two phenomena.

Fortunately it is possible in many cases to choose experimental conditions in such a manner that all but one or all but two broadening processes are either completely absent or negligibly small. For example, the absorption line produced in a continuous spectrum which passes through an attenuated beam of atoms moving perpendicular to the path of light would (if it could be spectroscopically resolved) be determined entirely by natural broadening. In a gas or vapour that is not electrically excited, and whose pressure is kept below 0·01 mm., Stark-effect broadening and Holtsmark broadening may be ignored. Lorentz broadening, that due to collisions with foreign gas molecules, however, cannot be disposed of so easily. In many experiments on resonance radiation one cannot get along without the use of foreign gases. In such cases, if the foreign gas pressure is kept below about 5 mm., the contribution to the absorption line due to Lorentz broadening is small in comparison with the Doppler effect. Lorentz broadening as a phenomenon in itself will be discussed later on in the book.

Most of the experiments leading to values of f or τ are performed under conditions in which only natural broadening and Doppler broadening are present. To interpret such experiments it is necessary to have a mathematical expression for the absorption coefficient of a gas under these conditions.

Such an expression was developed in 1912 by Voigt[75], and a little later independently by Reiche[60], on the basis of the classical electron theory. Voigt's formula is very general, involving the Doppler effect, natural damping, and any other damping process that can be represented by a function of the velocity of the absorbing atoms. When only the first two processes are considered Voigt's formula becomes identical with a formula which will be developed in the next few pages without the necessity of going through the long and somewhat complicated calculation of the dispersion theory. A fuller discussion of Voigt's and Reiche's formulas will be found in the Appendix.

It is a well-known result that, when natural damping is neglected, and only the heat motions of the atoms are taken into account, the absorption coefficient of a gas is given by

$$k_\nu = k_0 e^{-\left[\frac{2(\nu-\nu_0)}{\Delta\nu_D}\sqrt{\ln 2}\right]^2} \qquad \text{......(32)},$$

where $\Delta\nu_D$ is the Doppler breadth, depending only on the absolute temperature T and the molecular weight M according to the formula

$$\Delta\nu_D = \frac{2\sqrt{2R\ln 2}}{c}\nu_0\sqrt{\frac{T}{M}} \qquad \text{......(33)},$$

and k_0 is the *purely ideal quantity, the maximum absorption coefficient when Doppler broadening alone is present.* k_0 can be calculated as follows: Integrating Eq. (32), one obtains the formula

$$\int_0^\infty k_\nu d\nu = \frac{1}{2}\sqrt{\frac{\pi}{\ln 2}}k_0\Delta\nu_D \qquad \text{......(34)},$$

whereas Eqs. (28) and (29) yield

$$\int k_\nu d\nu = \begin{cases} \dfrac{\lambda_0{}^2 g_2}{8\pi g_1}\cdot\dfrac{N}{\tau}, \\[2ex] \dfrac{\pi e^2}{mc}\cdot Nf. \end{cases}$$

$$\text{Consequently} \quad k_0 = \begin{cases} \dfrac{2}{\Delta \nu_D} \sqrt{\dfrac{\ln 2}{\pi}} \cdot \dfrac{\lambda_0{}^2 g_2}{8\pi g_1} \cdot \dfrac{N}{\tau} \\[2ex] \dfrac{2}{\Delta \nu_D} \sqrt{\dfrac{\ln 2}{\pi}} \cdot \dfrac{\pi e^2}{mc} \cdot Nf \end{cases} \quad \ldots\ldots(35).$$

When, on the other hand, the Doppler effect is neglected, and only natural damping is taken into account, the absorption coefficient is proportional to

$$\frac{1}{1 + \left[\dfrac{2(\nu - \nu_0)}{\Delta \nu_N}\right]^2},$$

where $\Delta \nu_N$ is the natural breadth, which according to Dirac's theory of radiation [25, 82] is equal to the Einstein A coefficient divided by 2π or, in the case of a resonance line,

$$\Delta \nu_N = \frac{1}{2\pi\tau} \qquad \ldots\ldots(36).$$

Now the Doppler effect and natural damping are entirely independent broadening processes. Consequently the combined absorption coefficient of a gas (i.e. when both processes are present) may be calculated by considering either every infinitesimal frequency band of the pure Doppler curve to be broadened by natural damping, or every infinitesimal frequency band of the natural damping curve to be broadened by the Doppler effect. Suppose we pick some frequency band at a distance $\nu - \nu_0$ from the centre of a line showing only natural broadening. To represent the Doppler broadening of this frequency band, a variable distance δ from the point $\nu - \nu_0$ is chosen. The integration is then taken over δ. The absorption coefficient is therefore given by

$$k_\nu = C \int_{-\infty}^{\infty} \frac{e^{-\left[\frac{2\delta}{\Delta\nu_D}\sqrt{\ln 2}\right]^2}}{1 + \left[\dfrac{2}{\Delta\nu_N}(\nu - \nu_0 - \delta)\right]^2} \, d\delta \quad \ldots\ldots(37),$$

where C is a constant determined by the condition [Eq. (28)] that

$$\int_0^{\infty} k_\nu \, d\nu = \frac{\lambda_0{}^2 g_2}{8\pi g_1} \cdot \frac{N}{\tau}.$$

Integrating Eq. (37) with respect to ν, and using Eqs. (28) and (35), C is found to be

$$\frac{2k_0}{\pi \Delta \nu_N}.$$

The Doppler breadth of an absorption line offers itself as a convenient natural unit with which to describe an absorption line. Considerable simplification is therefore attained if the following two quantities are introduced:

$$\omega = \frac{2(\nu - \nu_0)}{\Delta \nu_D}\sqrt{\ln 2} \qquad \ldots\ldots(38),$$

$$a = \frac{\Delta \nu_N}{\Delta \nu_D}\sqrt{\ln 2} \qquad \ldots\ldots(39).$$

Letting $y = \frac{2\delta}{\Delta \nu_D}\sqrt{\ln 2}$, Eq. (37) becomes

$$k_\nu = k_0 \frac{a}{\pi}\int_{-\infty}^{\infty}\frac{e^{-y^2}dy}{a^2 + (\omega - y)^2} \qquad \ldots\ldots(40),$$

TABLE X

Atom	Line	λ_0	τ sec.	$\Delta \nu_N = \dfrac{1}{2\pi\tau}$ sec.$^{-1}$	$\Delta \nu_D = 7{\cdot}16 \times 10^{-7}\nu_0\sqrt{\dfrac{T}{M}}$ sec.$^{-1}$	$a = \dfrac{\Delta \nu_N}{\Delta \nu_D}\sqrt{\ln 2}$
Hg	$6\,^1S_0 - 6\,^3P_1$	2537	$1{\cdot}1 \times 10^{-7}$	$1{\cdot}4 \times 10^6$	$1{\cdot}0 \times 10^9$ (20° C.)	·0012
Na	$3\,^2S_{1/2} - 3\,^2P_{1/2}$	5896	$1{\cdot}6 \times 10^{-8}$	$1{\cdot}0 \times 10^7$	$1{\cdot}6 \times 10^9$ (160° C.)	·0052
Cd	$5\,^1S_0 - 5\,^1P_1$	2288	$2{\cdot}0 \times 10^{-9}$	$0{\cdot}8 \times 10^8$	$1{\cdot}9 \times 10^9$ (200° C.)	·035

which is identical with Voigt's expression (see Appendix). The quantity a will be called hereafter the "*natural damping ratio*". Since it is a constant for a particular absorption line of a gas at constant temperature, the integral in Eq. (40) is therefore a function of ω. Values of the natural damping ratio, a, for three important resonance lines are given in Table X. It is seen that a is always small, in the neighbourhood of 0·01. It is therefore of value to study the characteristics of an absorption line with a small natural damping ratio.

2b. CHARACTERISTICS OF AN ABSORPTION LINE WITH SMALL NATURAL DAMPING RATIO. It is shown in the Appendix that, when a is of the order of 0·01, Eq. (40) can be put in the form

$$\frac{k_\nu}{k_0} = e^{-\omega^2} - \frac{2a}{\sqrt{\pi}} [1 - 2\omega F(\omega)] \qquad \ldots\ldots(41),$$

where

$$F(\omega) = e^{-\omega^2} \int_0^\omega e^{y^2} \, dy \qquad \ldots\ldots(42).$$

A table of values of $F(\omega)$ and of $1 - 2\omega F(\omega)$ is given in the Appendix. With the aid of this table, the ratio k_ν/k_0 can be evaluated for any desired value of ω, that is, at any part of an absorption line whose natural damping ratio is known. In Table XI are given values of k_ν/k_0 from the centre of a line $(\omega = 0)$ to a distance about eight times the Doppler breadth $(\omega = \pm 16)$ for an absorption line whose natural damping ratio is $0 \cdot 01 \dfrac{\sqrt{\pi}}{2} = 0 \cdot 00886$, a value within the range of most resonance lines. It is of advantage to study this table in two parts: the "central region of the line", $|\omega| < 2$; and the "edges of the line", $|\omega| > 6$.

2c. THE CENTRAL REGION OF THE LINE. The fraction of the incident light of frequency ν that passes through an absorbing layer of thickness l is equal to $e^{-k_\nu l}$. Since $k_\nu l = \dfrac{k_\nu}{k_0} . k_0 l$, it is necessary to know the product $k_0 l$ in order to calculate how much light is transmitted at some particular part of an absorption line. From Eq. (35) it is evident that k_0 depends upon N, or more simply upon the pressure of the absorbing gas. If the pressure of the absorbing gas and the thickness of the absorbing layer are chosen low enough, $k_0 l$ can be made small, say about 3. (This is usually achieved at pressures from 10^{-7} to 10^{-4} mm., and with thicknesses from 0·1 to 3 cm.) In this case it is seen from Table XI that $k_\nu l$ has a value large enough to produce measurable absorption only within the central region of the line, being negligibly small for the values of $|\omega|$ greater than two. Moreover, it can be seen that the values of k_ν/k_0 in this part of the line differ from the values of $e^{-\omega^2}$ by only a few per cent. at most. With an error well within that of experiment the statement can be made: *when the pressure of the absorbing gas* *and the thickness of the absorbing layer are chosen small enough to make $k_0 l$ about 3, the edges of the absorption line may be neglected and the whole line may be regarded as a pure Doppler line with*

$$k_\nu = k_0 e^{-\omega^2} \qquad \ldots \ldots (43).$$

TABLE XI

ω	$e^{-\omega^2}$	$1 - 2\omega F(\omega)$ from Appendix	k_ν/k_0 from Eq. (41)	$k_0 l = 3$		$k_0 l = 3000$	
				$k_\nu l$	$e^{-k_\nu l}$	$k_\nu l$	$e^{-k_\nu l}$
·0	1·0000	1·0000	·9900	2·970	·0513		0
·2	·9608	·9221	·9516	2·855	·0576		0
·4	·8521	·7121	·8450	2·535	·0793		0
·6	·6977	·4303	·6934	2·080	·1249		0
·8	·5273	·1487	·5258	1·577	·2066	Too large to produce measurable transmission	0
1·0	·3679	− ·07616	·3687	1·106	·3309		0
1·2	·2369	− ·2175	·2391	·717	·4882		0
1·4	·1409	− ·2782	·1437	·431	·6499		0
1·6	·07730	− ·2797	·08010	·240	·7866		0
1·8	·03916	− ·2485	·04165	·124	·8834		0
2·0	·01832	− ·2052	·02037	·061	·9408		0
3	·0001234	− ·06962	·0008196	1		2·459	·0855
4	0	− ·03480	·0003480	1		1·044	·3520
5	0	− ·02134	·0002134	1		·6402	·5273
6	0	− ·01451	·0001451	1		·4353	·6473
7	0	− ·01053	·0001053	1	Too small to produce measurable absorption	·3159	·7291
8	0	− ·008000	·00008000	1		·2400	·7866
9	0	− ·006290	·00006290	1		·1887	·8278
10	0	− ·005076	·00005076	1		·1523	·8590
11	0	− ·004183	·00004183	1		·1255	·8816
12	0	− ·003510	·00003510	1		·1053	·9003
13	0	− ·002958	·00002958	1		·08874	·9148
14	0	− ·002551	·00002551	1		·07653	·9259
15	0	− ·002222	·00002222	1		·06667	·9352
16	0	− ·001953	·00001953	1		·05859	·9427

$2d$. THE EDGES OF THE LINE. If the pressure of the absorbing gas is from about 10^{-4} to about 10^{-2} mm. (not high enough to produce Holtsmark broadening of the absorption line!), and the thickness of the absorbing layer is from 10 to 50 cm., $k_0 l$ may be made very large, say about 3000. It is seen from the table in this case that $k_\nu l$ in the central region of the line is so large that, in any experiment, this region would be completely absorbed. Only in the edges of the line would a measurable amount of light be transmitted, that is, *the form of the line would be determined entirely by the edges*. At the edges of the line Eq. (41) assumes a very simple form. It is shown in the Appendix that, for large values of ω,

$$1 - 2\omega\, F(\omega) = -\left[\frac{1}{2\omega^2} + \frac{1.3}{(2\omega^2)^2} + \frac{1.3.5}{(2\omega^2)^3} + \frac{1.3.5.7}{(2\omega^2)^4} + \ldots\right] \ldots (44),$$

whence, for values of $|\,\omega\,|$ greater than 6, Eq. (41) becomes

$$k_\nu = \frac{a}{\sqrt{\pi}} \cdot \frac{k_0}{\omega^2} \qquad \ldots\ldots (45).$$

Introducing the values of ω and a given by Eqs. (38) and (39), Eq. (45) becomes

$$k_\nu = \left[\frac{1}{2}\sqrt{\frac{\pi}{\ln 2}}\, k_0 \Delta\nu_D\right] \frac{\Delta\nu_N}{2\pi\,(\nu - \nu_0)^2} \qquad \ldots\ldots (46),$$

and, by virtue of Eq. (34), Eq. (46) assumes the form

$$k_\nu = [\textstyle\int k_\nu\, d\nu]\, \frac{\Delta\nu_N}{2\pi\,(\nu - \nu_0)^2} \qquad \ldots\ldots (47),$$

which expresses the very interesting result that *the extreme edges of the line are due entirely to natural damping.*

The results of these calculations may be summed up as follows: When there is weak absorption the central region of the line plays the main rôle and the absorption coefficient is determined by the Doppler effect; whereas when there is very strong absorption, the edges of the line are important and the absorption coefficient is determined by damping. This is shown graphically in Figs. 22 and 23, in which the numbers in Table XI are plotted.

A very important distinction between the two cases arises when we consider the problem of hyperfine structure. The whole discussion up to now has been concerned with a simple line. When a line shows hyperfine structure, it may be regarded as being composed of a number of simple lines that are either completely separated (resolved) or overlap one another. All the formulas that have been written, therefore, must be applied to *each* hyperfine structure component. It is obvious then, that, in the case of weak absorption where the central region of each hyperfine structure component plays the main rôle, it is necessary to know the number of components, their separation, and their respective intensities, in order to give an exact expression for the absorption coefficient. In the case of strong absorption, however, where one is interested in the

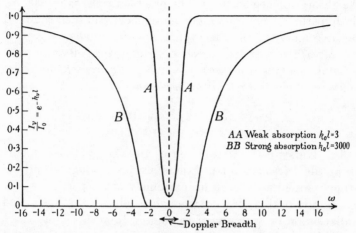

Fig. 22. Narrow and broad absorption lines.

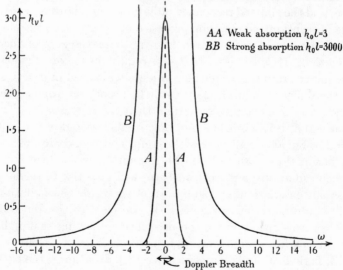

Fig. 23. Variation of absorption coefficient with frequency in narrow and broad absorption lines.

extreme edges of a line, at a distance from the centre of gravity of the hyperfine structure components of a line that is much larger than the average separation of the components, the situation is much simpler, for if this distance be large enough, the absorption coefficient will be practically the same as if all the hyperfine structure components coincided at their common centre of gravity. In other words, *in order to interpret experiments performed on the extreme edges of a line that has a complicated hyperfine structure it is a sufficiently accurate procedure to use equations that refer to a simple line.*

3. EMISSION AND DIFFUSION OF RESONANCE RADIATION

3*a*. EMISSION CHARACTERISTICS OF A RESONANCE LAMP. In an ideal resonance lamp there are just enough absorbing atoms present to absorb an extremely small portion of the exciting radiation, but not enough to absorb the re-emitted radiation on its way out of the lamp. In such a lamp there would be a uniform distribution of excited atoms in the direct path of the exciting radiation and no excited atoms anywhere else, and the emitted resonance radiation would be due to only one absorption and emission process on the part of each atom. Such resonance radiation is known as primary resonance radiation. It is obvious that the requirements of an ideal resonance lamp can never be completely satisfied in practice. If there are enough atoms present to absorb an appreciable amount of the exciting radiation, then the re-emitted radiation will also be absorbed, not only on its way out of the exit window but also in all parts of the lamp. The primary resonance radiation thus absorbed will be re-emitted (secondary resonance radiation), and this, in turn, will give rise to tertiary resonance radiation, and so on. In other words the radiation will be diffused or imprisoned. On the basis of the Einstein theory, this situation can be described by saying that light quanta perform rapid transits from atom to atom alternating with periods of imprisonment of average duration τ, the time τ being large compared with the time of transit. It was first pointed out by K. T. Compton[2, 3] that the propagation of resonance radiation in an absorbing gas was analogous to the process of diffusion, and, on the basis of this analogy, he was able to explain some phenomena in connection with the behaviour of low voltage arcs. A rigorous treatment of the problem was given later by E. A. Milne [48], who showed that the differential equations giving the concentration of excited atoms and the intensity of resonance radiation as a function of distance and time were similar to the ordinary diffusion equation except for a third order term that arose from the finite lifetime of the excited state. Milne's theory is of value in considering the one-dimensional flow of resonance radiation of very narrow

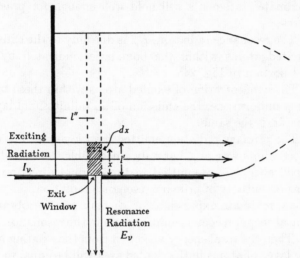

Fig. 24. Characteristics of an ideal resonance lamp.

spectral width, but leads to mathematical difficulties when applied to the more complicated conditions that are present in connection with an actual resonance lamp. The use of Milne's theory in interpreting experiments on the quenching of resonance radiation will be discussed in Chap. IV. At present an approximate treatment of the emission characteristics of a resonance lamp will be given in which the conditions of gas pressure and geometry are such that radiation diffusion may be neglected as a first approximation.

Let us consider the resonance lamp shown diagrammatically in Fig. 24. The exciting radiation I_ν passes down the lamp grazing the exit window, so that there is practically no layer of unexcited atoms through which the resonance radiation must pass on its way out. The aperture in the exit diaphragm is chosen small, and a lens is imagined at sufficient distance from the exit window to receive only a parallel beam of resonance radiation E_ν emerging perpendicularly from the exit window. Furthermore, the pressure of the absorbing gas in the resonance lamp is considered small enough, so that secondary and tertiary resonance radiation can be neglected as a first approximation. If these requirements are satisfied, then the following approximate statements will hold well enough for practical purposes:

(a) The resonance radiation, E_ν, is due only to the emission by excited atoms within the boundaries indicated by the shaded portion in Fig. 24.

(b) The concentration of excited atoms within these boundaries is uniform; i.e. the emission of all infinitesimal layers, such as dx, is the same.

(c) The primary radiation emitted by a layer such as dx is absorbed on its way out, but only a negligible fraction of this absorbed energy is re-emitted in the original direction; i.e. E_ν consists entirely of primary resonance radiation.

An approximate expression for E_ν may now be obtained. Let the absorption coefficient of the gas in the resonance lamp be k_ν'. Then the total energy absorbed from the exciting beam by the layer of atoms in the shaded area will be equal to

$$\text{const.} \int I_\nu e^{-k_\nu' l''} k_\nu' \, d\nu \qquad \ldots\ldots(48).$$

Some of this energy is re-emitted at right angles to the direction of the exciting beam. The experiments of Orthmann and Pringsheim[55] and of Rump[63] have shown that the *form* or frequency distribution of resonance radiation emitted at right angles to the direction of the exciting beam is independent of the form of the exciting line, and depends only on the form of the absorption coefficient. Therefore the radiation in a unit frequency band at the frequency ν emitted at right angles by the infinitesimal layer dx is proportional to

$$[\text{const.} \int I_\nu e^{-k_\nu' l''} k_\nu' \, d\nu] \,.\, k_\nu' \, dx = C k_\nu' \, dx \quad \ldots\ldots(49),$$

in which C will be constant, when the exciting lamp and the resonance lamp are run under constant conditions of temperature, pressure, etc. The radiation emitted by dx is absorbed on its way out, so that the amount emerging is

$$- C k_\nu' dx e^{-k_\nu' x}.$$

The emerging radiation from the whole layer of emitting atoms is therefore

$$E_\nu = - C k_\nu' \int_0^{l'} e^{-k_\nu' x} dx,$$

or
$$E_\nu = C (1 - e^{-k_\nu' l'}) \qquad \ldots\ldots(50).$$

This is then the expression for the frequency distribution of the radiation emitted by a resonance lamp approximating ideal conditions, and was first used by Ladenburg and Reiche [35] in connection with the emission characteristics of sodium flames. It is rigorously true only for an emitting layer of vanishingly small cross-section, that is, for a filament of length l', but can be expected to give a satisfactory description of the radiation from any resonance lamp in which the gas pressure is low enough to render the amount of secondary and tertiary resonance radiation small in comparison with the primary resonance radiation.

Other expressions have been used with varying degrees of success to represent the radiation emitted by a resonance lamp. Kunze [33] used a resonance lamp in which the distance l' was so small that Eq. (50) reduced to

$$E_\nu = \text{const.} \, k_\nu',$$

and then corrected this expression to take account of a small amount of absorption by a layer of unexcited atoms lying next to the exit window. Schein [65] and Zemansky [89] recognized that a line emitted by a resonance lamp is broader than the absorption line; and, in working with mercury resonance radiation under conditions in which $k_\nu' = k_0' e^{-\omega^2}$, used for the form of the emitted line an expression of the type $e^{-(\omega/\alpha)^2}$ which, with $\alpha > 1$, has a breadth greater than the curve $e^{-\omega^2}$. Although this is not as good an approximation as Eq. (50), it is still serviceable in many cases, particularly those in which there may be some doubt as to the applicability of Eq. (50).

3*b*. METHODS OF MEASURING LIFETIME. It is obvious that, for a resonance lamp which justifies the use of Eq. (50), a measurement of the total emitted radiation, $\int E_\nu d\nu$, is an indication of the number of excited atoms in the emitting layer. (This is strictly true only when E_ν is primary resonance radiation. If radiation diffusion plays a rôle, Milne showed that not only the number of excited atoms but also their gradient must be taken into account.) If such a resonance lamp be excited for a while, and the excitation then cut off, the decay of the re-emitted radiation, $\int E_\nu d\nu$, is a measure of the decay of the excited atoms. From the definition of the Einstein A coefficient the number of excited atoms n present at a time t after the cut-off of the excitation is given by

$$n = n_0 e^{-A_{2 \to 1} t} = n_0 e^{-t/\tau} \qquad \ldots\ldots (51),$$

provided

(*a*) the excited state is the resonance state,

(*b*) there is no further rate of formation of excited atoms due to radiation imprisonment, or to transitions from higher levels down to the resonance level.

If the above conditions are satisfied, an experimental curve of $\int E_\nu d\nu$ against the time should be exponential, with an exponential constant equal to $1/\tau$. In order to use Eq. (51) to measure τ, one must have a device for producing excited atoms that can be cut off suddenly, and a receiving device for measuring the radiation emitted after the cut-off. Several methods have been used, as follows:

3*c*. RESONANCE LAMP WITH ELECTRICAL CUT-OFF. In a tube first described by Webb [78] there is a filament-grid system, the gas to be investigated, and a grid-photoelectric plate system. An alternating potential between the filament and the grid produces an intermittent stream of electrons with sufficient energy to excite the atoms to the resonance state, and an alternating potential of the same frequency between the photoelectric plate and grid enables the resulting radiation to be measured during the proper half-cycle. By varying the frequency, the time between excitation and reception of the emitted radiation is varied, and from a curve of photoelectric current against frequency, the lifetime of the radiation may be calculated. In this way, Webb and Messenger[79] measured the lifetime of the resonance radiation of Hg, 2537, at low vapour pressure and found the limiting value to be about 10^{-7} sec.; and Slack[69] found the lifetime of the hydrogen resonance line, 1216, to be $1·2 \times 10^{-8}$ sec. With an improved apparatus of this type, in which the photoelectric system occupied a separate tube, Garrett[12] made a very careful investigation of the lifetime of mercury resonance radiation at very low vapour pressures and obtained for τ, $1·08 \times 10^{-7}$ sec., which is one of the most reliable values up to date.

$3d$. RESONANCE LAMP WITH OPTICAL CUT-OFF. The electro-optical phosphoroscope was first described by Abraham and Lemoine[1] and later used by Pringsheim to measure the velocity of light. In the modified form, due to Gaviola[13, 14], in which it can be used to measure the after-glow of radiation, it consists essentially of the following parts: (1) a beam of exciting light that is rendered intermittent by passing through a Kerr cell which is placed between crossed Nicols and across which is established an alternating potential; (2) a resonance lamp containing the gas to be studied; (3) another Kerr cell between crossed Nicols which allows the re-emitted resonance radiation to pass through intermittently because of the alternating potential on the Kerr cell. The use of this apparatus in order to measure the lifetime of the resonance radiation emitted by the gas in the resonance lamp is rather complicated, and for these details the reader is referred to the paper of Gaviola quoted above, and to a recent paper by Duschinsky[5a]. With this device Hupfield[27] measured the lifetime of Na resonance radiation consisting of both D lines. The Na vapour was at a pressure corresponding to a temperature between $190°$ C. and $200°$ C. at which some imprisonment of resonance radiation must have taken place. Unfortunately, the effect of vapour pressure on lifetime was not studied. The value obtained, however, $1·5 \times 10^{-8}$ sec., is in excellent agreement with those obtained by other methods to be described later. A repetition of this experiment, with substantially the same apparatus, was made by Duschinsky[5] in 1932 with startling results. First, the lifetime of the sodium resonance states was found to be $8 \cdot 2 \pm 3 \times 10^{-9}$ sec., almost one-half of Hupfield's value. Second, radiation imprisonment was found to begin at a temperature of 170° C., indicating that Hupfield's result (obtained at 190° C.) concerned the lifetime of the radiation caused by repeated emissions and absorptions, rather than the lifetime of the atom. Third, the presence of nitrogen shortened the lifetime, whereas the presence of helium did not. The lifetime $8 \cdot 2 \times 10^{-9}$ sec. corresponds to an f-value for both D lines of $1 \cdot 9$, whereas, according to the f sum rule, the sum of the f-values for *all* the lines of the principal series of an alkali atom should be equal to about 1. Moreover, measurements of magneto-rotation of polarized light at the edges of the resonance lines of Na, K and Cs yield f-values in good agreement with the theory. Duschinsky's result is therefore in very serious disagreement with both theory and experiment. It is further to be noted that, although there is no doubt that radiation imprisonment takes place in sodium vapour at a temperature higher than 170° C. (unpublished measurements of Zehden confirm this), there is still no assurance that the lifetime of the radiation measured under these conditions should be twice as large as the lifetime of the atom.

The dependence of the lifetime on the foreign gas pressure brings up the possibility of a new kind of collision process, namely, one in which an excited atom is stimulated by an impact with a foreign gas molecule to radiate sooner than it ordinarily would. If such a collision process takes place, it is hard to understand why a nitrogen molecule should be effective and a helium atom should not. In fact, one would expect the opposite to be the case, inasmuch as nitrogen is known to quench sodium resonance radiation, thereby taking the excitation energy of the excited sodium atom away and not allowing it to radiate, whereas helium does not quench sodium resonance radiation. An experiment on sodium resonance radiation was performed by von Hamos[20] to discover whether such a process takes place, and it was found that the results could be explained quite adequately in terms of collision broadening of the absorption line by the foreign gas molecules. This experiment, however (which is discussed more in detail in Chap. IV under "Lorentz Broadening"), does not prove conclusively that no such process exists. Moreover, other experiments seem to indicate the possibility of a "collision-stimulated" emission of an excited atom. It is therefore worth while, at this point, to consider the consequences of such a process in the light of thermodynamic equilibrium, to see if it is theoretically sound.

First of all, if there is to be thermodynamic equilibrium in an enclosure containing radiation, absorbing atoms and foreign gas molecules, a "collision-stimulated" spontaneous emission cannot be assumed without also assuming "collision-stimulated" absorption and "collision-stimulated" induced emission. That is, corresponding to the three Einstein coefficients defined early in this chapter (in capital letters), there must be three more, say, $b_{1 \to 2}$, $a_{2 \to 1}$ and $b_{2 \to 1}$ with the same definitions as the Einstein coefficients except that, instead of being atomic constants, they are all proportional to the foreign gas pressure. It can easily be shown that, upon introducing these three new coefficients, Planck's law can be derived.

Now, if the "collision-stimulated" absorption and "collision-stimulated" induced emission be added to the Einstein absorption and induced emission in the derivation in §1b, Eq. (28) becomes

$$\int k_\nu d\nu = \frac{\lambda_0^2 g_2}{8\pi g_1} . N \left(a_{2 \to 1} + A_{2 \to 1}\right)\left(1 - \frac{g_1}{g_2}\frac{N'}{N}\right)\dots\dots(52),$$

which shows that, if the number of atoms in the normal state, N, remain constant, and the ratio N'/N (fraction of excited atoms) is small, the integral of the absorption coefficient should *increase* with the foreign gas pressure, since $a_{2 \to 1}$ is proportional to the pressure. The integral $\int k_\nu d\nu$ over the mercury resonance line 2537 was measured by Füchtbauer, Joos and Dinkelacker[11] in the presence of foreign gases at pressures from 10 to 50 atmospheres and, instead of finding an increase, they noted a *decrease* of at most about 20 per cent. If one could be certain that the factor $N\left(1 - \dfrac{g_1}{g_2}\dfrac{N'}{N}\right)$ remained constant during the experiment, one might conclude from this that the presence of a "collision-stimulated" emission on the part of an excited atom was disproved. No such conclusion, however, can be made, because there is a possibility that $N\left(1 - \dfrac{g_1}{g_2}\dfrac{N'}{N}\right)$ became smaller as the foreign gas pressure increased, thereby masking any increase due to $a_{2\to1}$. The possibility of a decrease in $N\left(1 - \dfrac{g_1}{g_2}\dfrac{N'}{N}\right)$ arises from the fact that the metastable $6\,^3P_0$ level of mercury lies so close to the radiating state $6\,^3P_1$. If metastable atoms are formed by collision and are prevented from diffusing to the walls by the tremendous foreign gas pressure, they will then be raised again to the $6\,^3P_1$ state, and mercury atoms will continually be oscillating between the two states. This will result in a high population of the $6\,^3P_0$ state and a consequent reduction in N. This is given only as a *possible* explanation of Füchtbauer's results, in order to show that they do not completely preclude the possibility of a collision-stimulated emission. The whole question must, at this time, be left open until further experimental work is done.

3 e. ATOMIC RAY OPTICALLY EXCITED. Perhaps the most direct measurement of the lifetime of a resonance state consists in the measurement of the light emitted by a beam of atoms moving perpendicularly through a narrow beam of exciting radiation [4]. The distance along the beam measured from the point where it is illuminated by the exciting radiation is a measure of the time after excitation, and the intensity of the radiation emitted at some point on the atomic beam is proportional to the number of excited atoms at the point. The curve obtained by plotting intensity of radiation against time after excitation is exponential, with an exponential constant equal to $1/\tau$ in the case of resonance radiation. This method is applicable only to an atom whose lifetime is of the order of 10^{-6} sec. or more, because the maximum thermal velocity

obtainable in experiments of this sort is of the order of 10^5 cm./sec. Koenig and Ellett[30] used this method to measure the lifetime of the $5\,^3P_1$ state of the cadmium atom by sending a beam of cadmium atoms perpendicularly through a narrow beam of radiation of wave-length 3261 (5^1S_0–$5\,^3P_1$), and found τ to be $2\cdot5 \times 10^{-6}$ sec. A qualitative experiment of the same sort was performed by Soleillet[71] on the same cadmium line, confirming Koenig and Ellett's result.

$3f$. CANAL RAY. In order to measure somewhat shorter lifetimes than 10^{-6} sec., a beam of ions (canal ray) accelerated by an electric field to any desired velocity may be used. As the ions recombine and emit light, the intensity along the beam can be used to show the decay of the number of resonance atoms, provided sufficient time has elapsed for transitions from higher levels to the resonance level to take place. This method then leads to accurate results only when the lifetime of the resonance state is longer than that of higher states. Wien[84], Kerschbaum[28, 29] and Rupp[64] have used this method extensively with Hg, Na, H, Li, K, Ca and Sr, but only with moderate success. Wien's value of τ for the $6\,^3P_1$ state of mercury, $0\cdot98 \times 10^{-7}$ sec., agrees well with other measurements, but all other values are much too large—a result that is due undoubtedly to the presence of transitions from higher atomic levels. It is an important point that, in all cases, the decay curves appeared to be exponential, although, on account of transitions from higher levels, the curves would have to be represented by a sum of exponentials each with different exponential constants. That such a series can appear exponential, and yet yield a value of the exponential constant far from the actual value, is a danger that must be guarded against in all work in which the lifetime of a particular level is to be inferred from the decay of a spectral line.

$3g$. ABSOLUTE INTENSITY OF A RESONANCE LINE. If a gas is in thermal equilibrium at the temperature T, then the number of excited atoms per c.c., n, will be given by

$$\frac{n}{N} = \frac{g_1}{g_2} e^{-\epsilon/kT},$$

where N is the number of normal atoms per c.c., g_1 and g_2 are the statistical weights of the normal and the excited states respectively, and ϵ the energy difference between the normal and excited states. If the excited state be the resonance state, the total energy E emitted by the excited atoms will be

$$E = \frac{n}{\tau} h\nu = \frac{N h\nu}{\tau} \frac{g_1}{g_2} e^{-\epsilon/kT} \qquad \ldots\ldots(53).$$

A measurement, therefore, of the absolute intensity of a resonance line, E, enables one to measure τ. The first measurements of this kind were made by Gouy [16, 17, 18] and by Zahn [86] on the sodium D lines, and were used by Ladenburg [36] to obtain an estimate of the lifetime of the resonance states. An early measurement of the emission of a sodium flame by Ornstein and van der Held [52] yielded the value 5×10^{-8} sec. for the lifetime of the $3\,^2\mathrm{P}$ states, which is more than three times as large as the accepted value. This discrepancy was explained by Ladenburg and Minkowski [40] as due to an error in estimating N from the degree of dissociation of the salt (Na_2CO_3) which was used in the flame. A repetition of this experiment by van der Held and Ornstein [23] in 1932 yielded the value $\tau = 1\cdot63 \times 10^{-8}$ sec., in good agreement with other results, and also showed that the discrepancy present in the earlier determination was due in part to the slow vaporization of the water droplets present in the Na_2CO_3 solution that was sprayed into the flame.

4. ABSORPTION WITHIN AND AT THE EDGES OF A RESONANCE LINE

4a. AREA UNDER THE ABSORPTION COEFFICIENT. It was proved in § 2 of this chapter that, whatever physical properties are responsible for the formation of a resonance absorption line, the following equation, Eq. (28), remains valid:

$$\int k_\nu \, d\nu = \frac{\lambda_0{}^2 g_2}{8\pi g_1} \cdot \frac{N}{\tau}.$$

If a continuous spectrum be passed through a gas, and the

transmitted light be measured as a function of the frequency, the absorption coefficient may be calculated and plotted against the frequency. By graphical integration, then, the integral $\int k_\nu d\nu$ may be obtained, and from a knowledge of N, τ of the resonance state may be calculated. The outstanding difficulty connected with this method of measuring τ is the narrowness of most resonance lines. It is obvious that, if the monochromatic image of the slit of a spectrograph on the photographic plate cover a frequency range larger than the spectral width of the absorption line, the distribution of blackening on the plate will give no indication whatever of the true form of the absorption line. It is therefore necessary to work only with very broad absorption lines if the spectrograph is to have the usual working slit-width. This may be done by introducing a foreign gas at a very high pressure and making use of Lorentz broadening to such an extent that the absorption line has a much larger width than the monochromatic image of the spectrograph slit on the photographic plate.

Using several foreign gases at pressures ranging from 1 to 50 atmospheres, Füchtbauer, Joos and Dinkelacker[11] measured $\int k_\nu d\nu$ for the mercury resonance line 2537 when the mercury vapour pressure corresponded to a temperature of 18° C., and found that the value of the integral decreased with increasing foreign gas pressure. Extrapolating to zero foreign gas pressure, Tolman[74] showed that the integral yielded a value of τ equal to $1 \cdot 0 \times 10^{-7}$ sec. The decrease in $\int k_\nu d\nu$ with foreign gas pressure has already been commented on, and a possible explanation in terms of metastable atoms has been suggested. There remains to be pointed out only the fact that the extrapolation to zero foreign gas pressure is decidedly necessary in order to obtain a good value of τ, and that this extrapolation constitutes the main error in this method which otherwise is fairly simple and direct.

4*b*. ABSORPTION COEFFICIENT AT THE CENTRE OF A RESONANCE LINE. If resonance radiation from a resonance lamp be passed through a narrow absorption cell containing the same gas that is in the resonance lamp, and the ratio of the transmitted to the incident radiation be measured, the "absorption" A can be calculated as follows:

$$A = 1 - \frac{\text{Transmitted radiation}}{\text{Incident radiation}} \qquad \ldots \ldots (54).$$

If the frequency distribution of the incident radiation (the radiation emitted by the resonance lamp) is denoted by E_ν and the absorption coefficient of the gas in the absorption cell by k_ν, and the thickness of the absorption cell by l, the absorption is given by

$$A = 1 - \frac{\int E_\nu e^{-k_\nu l} d\nu}{\int E_\nu d\nu},$$

or

$$A = \frac{\int E_\nu (1 - e^{-k_\nu l}) d\nu}{\int E_\nu d\nu} \qquad \ldots \ldots (55).$$

There are two important applications of the above equation, as follows:

4c. METHOD OF LADENBURG AND REICHE. Suppose that the source of resonance radiation is a resonance lamp of the type shown in Fig. 24, containing a gas whose absorption coefficient is k_ν' and with an emitting layer of thickness l'. Then from Eq. (50) $E_\nu = C(1 - e^{-k_\nu' l'})$. If the radiation from this lamp is sent through an absorption cell of thickness l, containing a gas whose absorption coefficient is k_ν, then from Eq. (55) the absorption is given by

$$A = \frac{\int (1 - e^{-k_\nu' l'})(1 - e^{-k_\nu l}) d\nu}{\int (1 - e^{-k_\nu' l'}) d\nu} \qquad \ldots \ldots (56).$$

Ladenburg and Reiche[35] considered an experimental situation in which the emitting layer of the resonance lamp and the absorbing layer in the absorption cell *were identical in every respect*, i.e. in temperature, vapour pressure and thickness. This provides that $k_\nu' l' = k_\nu l$, whence, calling the absorption A_L in this case,

$$A_L = \frac{\int (1 - e^{-k_\nu l})^2 d\nu}{\int (1 - e^{-k_\nu l}) d\nu} \qquad \ldots \ldots (57).$$

The quantity A_L has been called by Ladenburg and Reiche the "line-absorption" ("Linienabsorption"). Measurements of line-absorption are practicable only when the vapour pressure in both the resonance lamp and the absorption cell is low. If, in addition to this, the natural damping ratio of the absorption line is of the order of 0·01 or smaller (which has been shown to be the case with most resonance lines), then the absorption coefficient k_ν can be expressed by Eq. (43), namely

$$k_\nu = k_0 e^{-\omega^2},$$

and A_L becomes finally

$$A_L = \frac{\int_{-\infty}^{\infty} (1 - e^{-k_0 l \, e^{-\omega^2}})^2 \, d\omega}{\int_{-\infty}^{\infty} (1 - e^{-k_0 l \, e^{-\omega^2}}) \, d\omega} \qquad \dots\dots(58).$$

Eq. (58) has been evaluated by Kopfermann and Tietze[31], and also by Ladenburg and Levy[42]. A table of values of A_L for various values of $k_0 l$ is given in the Appendix, and A_L is plotted against $k_0 l$ in Fig. 25. It is evident from this curve that one can obtain the value of $k_0 l$ that corresponds to any experimentally observed value of A_L. From Eq. (35), τ or f is given by

$$\frac{1}{\tau} = \frac{\Delta\nu_D}{2} \sqrt{\frac{\pi}{\ln 2}} \cdot \frac{8\pi g_1}{\lambda_0^2 g_2} \cdot \frac{k_0}{N},$$

$$f = \frac{\Delta\nu_D}{2} \sqrt{\frac{\pi}{\ln 2}} \cdot \frac{mc}{\pi e^2} \cdot \frac{k_0}{N},$$

which enable one to calculate τ or f from a knowledge of k_0, N and $\Delta\nu_D$. The usual procedure is to measure A_L for various values of N, and to plot the resulting values of $k_0 l$ against Nl. The slope of the resulting straight line yields k_0/N, which is then used to calculate τ or f.

Measuring $k_0 l$ for various values of Nl by the method of Ladenburg and Reiche involves varying the conditions in *both* the resonance lamp and the absorption cell at the same time and in exactly the same way. This is not always convenient. From an experimental standpoint it may be desirable to keep the resonance lamp under strictly constant conditions of temperature, pressure and thickness of emitting layer (optimum conditions), and to vary the temperature, pressure and thickness of the absorption cell only. If the variations in temperature of the gas in the absorption cell are not large enough to cause the Doppler breadth to differ from that in the resonance lamp by more than a few per cent. (which is usually the case, since Doppler breadth varies as the square root of the absolute tem-

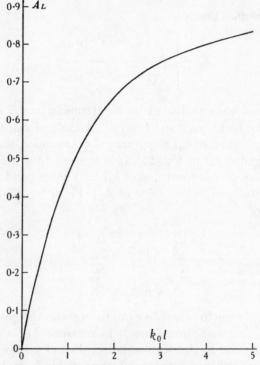

Fig. 25. Line absorption of Ladenburg and Reiche.

perature), the absorption coefficients of the gas in the absorption cell and in the resonance lamp may be written, respectively,

$$k_\nu = k_0 e^{-\omega^2},$$
$$k_\nu' = k_0' e^{-\omega^2},$$

where $\omega = \dfrac{2\,(\nu - \nu_0)}{\Delta\nu_D}\sqrt{\ln 2}$ and $\Delta\nu_D$ is the constant Doppler breadth of the gas in the resonance lamp. Calling the absorption in this case $A'_{k_0'l'}$, Eq. (56) becomes

$$A'_{k_0'l'} = \frac{\int_{-\infty}^{\infty} (1 - e^{-k_0'l'\,e^{-\omega^2}})\,(1 - e^{-k_0 l\,e^{-\omega^2}})\,d\omega}{\int_{-\infty}^{\infty} (1 - e^{-k_0'l'\,e^{-\omega^2}})\,d\omega} \quad ...(59).$$

Now, as the pressure of the absorbing gas in the cell is varied, there will, in general, be one value of Nl in the absorption cell which is identical with that in the resonance lamp. At this value of Nl, $k_0 l$ will be practically equal to $k_0'l'$ (neglecting again the slight difference between the two Doppler breadths), and this value of $k_0'l'$ may be found by measuring the absorption at this value of Nl and using the Ladenburg-Reiche curve in Fig. 25. Substituting this value of $k_0'l'$ in Eq. (59) there is obtained a relation between the absorption and $k_0 l$ from which the value of $k_0 l$ corresponding to *any* experimentally observed absorption (at *any* value of Nl) may be obtained. Eq. (59) has been evaluated for several values of $k_0'l'$ and $k_0 l$ and a table of values of $A'_{k_0'l'}$ is given in the Appendix.

$4d$. METHOD IN WHICH $E_\nu = Ce^{-(\omega/\alpha)^2}$. In order to employ the method of Ladenburg and Reiche it is necessary to have a resonance lamp which satisfies the conditions given on p. 108 and which is constructed according to Fig. 24. This is not always feasible. It is often necessary to interpret experiments on the absorption of light from a source excited by electron bombardment, or from a resonance lamp in which either the vapour pressure or the thickness of the emitting layer or both are not accurately known. In such cases it is convenient to use for the frequency distribution of the emitted radiation an empirical expression which represents roughly the line broadening resulting from the vapour pressure and temperature conditions within the lamp. A convenient expression for this purpose is

$$E_\nu = Ce^{-(\omega/\alpha)^2} \quad\quad ......(60),$$

where $\omega = \dfrac{2\,(\nu - \nu_0)}{\Delta\nu_D}\sqrt{\ln 2}$ and $\Delta\nu_D$ is the Doppler breadth of the *absorption line*. It is evident in this expression that

$$\alpha = \frac{\text{Emission line breadth}}{\text{Absorption line breadth}},$$

so that a value of α equal to unity implies a line of the same shape and breadth as that of the absorption coefficient of the gas in the absorption cell. A value of α greater than unity represents a line of the same shape but of greater breadth than that of the absorbing gas. In the case of a lamp with a very thin emitting layer, the expression $E_\nu = Ce^{-(\omega/\alpha)^2}$ represents rigorously the emitted radiation, and α is given by the square root of the ratio of the absolute temperatures of the emitting gas and the absorbing gas.

Setting k_ν equal to $k_0 e^{-\omega^2}$ as before, and calling the absorption in this case A_α, Eq. (55) becomes

$$A_\alpha = \frac{\int_{-\infty}^{\infty} e^{-(\omega/\alpha)^2} (1 - e^{-k_0 l e^{-\omega^2}})\, d\omega}{\int_{-\infty}^{\infty} e^{-(\omega/\alpha)^2}\, d\omega} \qquad \ldots\ldots(61),$$

which has been evaluated by Malinowski[45], Orthmann[54], Kunze[33], de Groot[19] and Zemansky[89]. A table of values of A_α for various values of α and of $k_0 l$ is given in the Appendix. To find α in any experiment in which the operating conditions of the lamp remain constant, one can proceed as follows: Choose any value of α, and use the resulting curve of A_α against $k_0 l$ to give the values of $k_0 l$ corresponding to experimentally measured values of the absorption. Plot these values of $k_0 l$ against Nl. According to Eq. (35), this should be a straight line. If it is not, choose another α until one is found that yields a straight line between $k_0 l$ and Nl. This value of α can then be used to describe approximately the radiation of the lamp in question in any further work that is done with the same lamp under the same operating conditions.

Although the three methods just described refer to an absorption line whose structure is simple, it is obvious that they may, with equal validity, be applied to an absorption line which consists of any number of *equal, completely separate, simple lines*.

4*e*. MEASUREMENTS ON SIMPLE LINES. The absorption of the mercury resonance line 2537 from a resonance lamp by a column of mercury vapour was first investigated in 1914 by Malinowski[45] who assumed that the radiation emitted by the resonance lamp had a pure Doppler form; i.e. he used Eq. (61) with $\alpha = 1$. Later measurements of Orthmann[54], Goos and Meyer[15], Hughes and Thomas[26], Thomas[73] and Schein[65] gave rather discordant results, partly because of insufficient control of experimental conditions and partly because of the inapplicability of the absorption formulas that were used. In recent years a number of new measurements of the absorption coefficient of the mercury resonance line have agreed well among themselves and have shown themselves to be consistent with a value of τ very close to 10^{-7} sec. In all of these experiments, the assumption was made that the line consists of five equal, completely separate lines, enabling the absorption formulas for a simple line to be used. It will be shown later that this is by no means a bad approximation.

Using the method of Ladenburg and Reiche, Kopfermann and Tietze[31] measured the absorption of mercury vapour and found the maximum absorption coefficient (i.e. the average absorption coefficient at the centre of the five components) to vary with the number of absorbing atoms, N, very nearly linearly for small values of N according to the law

$$k_0 = 1 \cdot 34 \times 10^{-13} N,$$

corresponding to a value of τ equal to $1 \cdot 05 \times 10^{-7}$ sec. Kunze[33] used a resonance lamp with a very thin emitting layer lying behind a thin non-emitting layer and used Eq. (61) with $\alpha = 1$ and with a correction term to take account of the absorption of the non-luminous vapour. He found that for small values of N, k_0 varied very nearly linearly with N such that

$$k_0 = 1 \cdot 38 \times 10^{-13} N,$$

corresponding to $\tau = 1 \cdot 04 \times 10^{-7}$ sec. Using Eq. (61) with $\alpha = 1 \cdot 21$, Zemansky's measurements[89] of the absorption of mercury resonance radiation yielded the result that

$$k_0 = 1 \cdot 41 \times 10^{-13} N,$$

which is consistent with a value of τ equal to $1 \cdot 0 \times 10^{-7}$ sec.

The absorption of the cadmium resonance line, 2288, was measured by Zemansky[90], using the modified method of Ladenburg and Reiche [Eq. (59)]. Since the line emitted from his resonance lamp showed practically no hyperfine structure, the absorption formula for a simple line could be used with confidence. The results obtained were: $k_0 = 1 \cdot 64 \times 10^{-11} N$ and $\tau = 2 \cdot 0 \times 10^{-9}$ sec.

4f. ABSORPTION OF A NUMBER OF SEPARATE SIMPLE LINES OF DIFFERENT INTENSITIES. If we denote the separate hyperfine-structure components by superscripts, the absorption $\bar{A}$ is given by the formula analogous to Eq. (55), as follows:

$$\bar{A} = \frac{\int E_\nu^{(1)}(1 - e^{-k_\nu^{(1)}l})\,d\nu + \int E_\nu^{(2)}(1 - e^{-k_\nu^{(2)}l})\,d\nu + \dots}{\int E_\nu^{(1)}\,d\nu + \int E_\nu^{(2)}\,d\nu + \dots} \quad \dots (62).$$

Using the method of Ladenburg and Reiche, with $k_\nu^{(i)} = k_0^{(i)} e^{-\omega^2}$,

$$\bar{A}_L = \frac{\displaystyle\int_{-\infty}^{\infty}(1 - e^{-k_0^{(1)}le^{-\omega^2}})^2\,d\omega + \int_{-\infty}^{\infty}(1 - e^{-k_0^{(2)}le^{-\omega^2}})^2\,d\omega + \dots}{\displaystyle\int_{-\infty}^{\infty}(1 - e^{-k_0^{(1)}le^{-\omega^2}})\,d\omega + \int_{-\infty}^{\infty}(1 - e^{-k_0^{(2)}le^{-\omega^2}})\,d\omega + \dots},$$

which is analogous to Eq. (58). If we write

$$\int_{-\infty}^{\infty}(1 - e^{-k_0 le^{-\omega^2}})\,d\omega = \sqrt{\pi}\,k_0 lS,$$

then, from Eq. (58),

$$\int_{-\infty}^{\infty}(1 - e^{-k_0 le^{-\omega^2}})^2\,d\omega = \sqrt{\pi}\,k_0 lSA_L,$$

and $\bar{A}_L$ becomes

$$\bar{A}_L = \frac{[k_0 lSA_L]^{(1)} + [k_0 lSA_L]^{(2)} + \dots}{[k_0 lS]^{(1)} + [k_0 lS]^{(2)} + \dots} \quad \dots \dots (63).$$

Both S and A_L are functions of $k_0 l$, and are tabulated in the Appendix for many values of $k_0 l$. In order, therefore, to calculate $\bar{A}_L$, it is merely necessary to know the respective $k_0 l$'s for all the hyperfine-structure components. These are determined as follows: If there are n hyperfine-structure components, then experimental or theoretical determinations of the intensities of the components yield the $n-1$ equations:

$$[k_0 l]^{(1)} : [k_0 l]^{(2)} : \dots = a : b : \dots,$$

where a, b, etc. are the intensities. The one more equation needed to compute all the $k_0 l$'s is supplied by Eq. (28), which becomes in this case:

$$\frac{1}{2}\sqrt{\frac{\pi}{\ln 2}}\,\Delta\nu_D\,\{[k_0 l]^{(1)}+[k_0 l]^{(2)}+\ldots\}=\begin{cases}\dfrac{\lambda_0{}^2 g_2}{8\pi\tau g_1}\cdot Nl,\\[2ex]\dfrac{\pi e^2 f}{mc}\cdot Nl.\end{cases}$$

The above equation is a result of an assumption that is justified theoretically, namely, that all hyperfine levels of a resonance level have the same lifetime. Knowing $\Delta\nu_D$ and either τ or f, all the $k_0 l$'s are determined as functions of Nl. $\bar{A}_L$ may then be calculated for a number of convenient values of Nl.

On the basis of Schüler and Keyston's analysis of the hyperfine structure of the mercury 2537 line (see Chap. I), Zehden and Zemansky [87] calculated by the above method the absorption $\bar{A}_L$ as a function of Nl. The resulting theoretical curve agreed exceedingly well with the experimental curve of Kopfermann and Tietze [31], provided τ was taken to be $1 \cdot 08 \times 10^{-7}$ sec. in agreement with Garrett's [12] value. On the basis of the simple picture of the hyperfine structure, i.e. five approximately equal components, the agreement between theory and experiment is fairly satisfactory (within 5 per cent.), so that this simple picture is still useful in cases where the consideration of the accurate hyperfine structure leads to too great a complication in calculating, and where the magnitude of the experimental error does not warrant refinements in calculation.

Using a mercury lamp with a very thin emitting layer excited by electron bombardment and operating under constant conditions, Garrett [12] measured the absorption of the mercury resonance line, 1849, $6\,^1S_0-6\,^1P_1$, by a column of mercury vapour contained in an absorption cell. From Schüler and Keyston's analysis of the mercury line, 4916, $6\,^1P_1-7\,^1S_0$, Garrett was able to deduce the fine structure of the 1849 line, and by the methods described above was able to interpret his experiments in terms of the absorption of five completely separate simple lines of different intensity. Since the lamp in this experiment had a very thin emitting layer, and since the square root of the ratio of the absolute temperature of the emitting gas to that of the absorbing gas was 1·2, Garrett used for each hyperfine-structure component of the emission line an expression of the type $E_\nu^{(i)} = k_0^{(i)} l e^{-\left(\frac{\omega}{1\cdot2}\right)^2}$. The absorption formula used to interpret the experiments was as follows:

$$\bar{A} = \frac{[k_0 l A_\alpha]^{(1)} + [k_0 l A_\alpha]^{(2)} + \dots}{[k_0 l]^{(1)} + [k_0 l]^{(2)} + \dots},$$

with $\alpha = 1\cdot2$. The experimental results were found to be consistent with a value for the lifetime of the $6\,^1P_1$ state of the mercury atom equal to $0\cdot3 \times 10^{-9}$ sec., the smallest value of τ that has yet been measured.

4g. ABSORPTION OF A LINE WITH OVERLAPPING COMPONENTS. When the hyperfine-structure components of a line overlap, the absorption must be calculated by graphical integration. Using the method of Ladenburg and Reiche, Eq. (57) can be evaluated graphically as soon as a graph of $k_\nu l$ as a function of frequency can be drawn. This is done as follows: First, the various $k_0 l$'s are calculated as in the preceding section. A number of Gauss error curves are then drawn side by side so that each curve has a maximum height equal to one of the $k_0 l$'s and a breadth equal to $\Delta\nu_D$. The separations between the curves are made equal to the measured hyperfine-structure separations. The curves are then added, and the resulting curve represents $k_\nu l$. The curves of $(1 - e^{-k_\nu l})$ and $(1 - e^{-k_\nu l})^2$ are then plotted against frequency and, by graphical integration, A_L is obtained.

The only case in which this procedure has been carried out carefully is that of the sodium resonance lines. The absorption of the sodium D lines emitted by a resonance lamp of the type shown in Fig. 24 was measured very accurately by Zehden[88] for various values of the sodium vapour pressure in the absorption cell. The experiment was performed by the method of Ladenburg and Reiche and the results were expressed as curves of absorption against Nl. In order to calculate the absorption A_L according to the method just outlined, it was necessary first to know the intensity ratios of the hyperfine-structure components of both the D lines. According to Schüler each line is a doublet with a separation of 0·060 cm.$^{-1}$, owing to the splitting up of the $^2S_{1/2}$ state, the splitting of the $^2P_{1/2}$ and $^2P_{3/2}$ states being too small to produce any effect. Since the nuclear moment of the sodium atom was not known at the time, and since there existed no reliable measurements of the intensity ratio of the components, it was necessary for Zehden to try several values for the nuclear spin. Assuming the spin to be $\frac{1}{2}$, and the lifetime of the ^2P states to be $1·6 \times 10^{-8}$ sec., a theoretical curve of A against Nl was obtained which did not agree satisfactorily with the experimental curve. Recent measurements of the separation of the hyperfine-structure components of each D line by Van Atta and Granath[74b], and of the nuclear spin (3/2) by Rabi and Cohen[59a], will enable Zehden's measurements to be re-interpreted when they are published.

4h. ABSORPTION OF A GAS IN A MAGNETIC FIELD. In order to obtain an expression for the absorption of resonance radiation by a gas in a magnetic field it is necessary merely to use for the absorption coefficient the sum of a number of Gauss error curves, with wave-length separations determined by the Zeeman effect, as was first shown by Malinowski[45]. The details of such a calculation are given in a paper by Schein[65], who derived the absorption formula applicable to the absorption of mercury resonance radiation by mercury vapour placed in a magnetic field. The calculations agreed approximately with the experimentally observed decrease of the absorption as the magnetic field was varied from 0 to 1000 gauss. The complete experimental curve of absorption against magnetic field up to 13,000 gauss showed five maxima.

By placing an absorption cell containing mercury vapour between the poles of an electromagnet, Mrozowski[51] was able to show that, at certain values of the field strength, only one or two hyperfine-structure components of the mercury 2537 line were transmitted. In this way he was able to investigate one component by itself and the other four components in combinations of two. By measuring the magnetic depolarization of the resonance radiation (see Chap. v) excited in mercury vapour by one or two hyperfine-structure components of the 2537 line, Mrozowski arrived at a value of τ for each kind of radiation. The three values of τ so obtained differed from one another, in disagreement with the ideas of Schüler and Keyston, whose analysis of the hyperfine structure of the mercury 2537 line indicates that the lifetimes of all the hyperfine levels are the same. A re-evaluation of Mrozowski's results by Mitchell (see Chap. v) indicates the same lifetime for all hyperfine levels, but that the absolute value of this lifetime is larger than the usually accepted value. There is a possibility, in spite of the author's assertion to the contrary, that the discrepancy is within the limits of experimental error.

4 i. ABSORPTION COEFFICIENT AT THE EDGES OF A RESONANCE LINE. It was shown in Sect. 2 d that the extreme edges of an absorption line are due entirely to natural damping and that the absorption coefficient very far from the centre of the line is given by Eq. (47), namely,

$$k_\nu = [\textstyle\int k_\nu d\nu] \cdot \frac{\Delta\nu_N}{2\pi\,(\nu - \nu_0)^2}.$$

Since
$$\int k_\nu d\nu = \frac{\lambda_0^{\,2} g_2}{8\pi g_1} \cdot \frac{N}{\tau} \quad \text{and} \quad \Delta\nu_N = \frac{1}{2\pi\tau},$$

$$k_\nu = \frac{\lambda_0^{\,2} g_2}{8\pi g_1} \cdot \frac{N}{\tau^2} \cdot \frac{1}{4\pi^2\,(\nu - \nu_0)^2},$$

and if k_ν, N and $(\nu - \nu_0)$ are measured, τ may be calculated from the above formula. This is not, however, the procedure that has been adopted in the past, for the reason that it has not always been possible to measure N either because of a lack of knowledge of the vapour pressure curve, or because non-uniform temperature conditions of the absorption tube prohibited the use of any vapour pressure data. To avoid the necessity of knowing N, it has been customary to perform two different experiments with the same apparatus, and to eliminate N between them. In order to explain the way this is done it is convenient to put Eq. (47) into the classical form in which it has most often been used.

Using the classical result

$$\int k_\nu d\nu = \frac{\pi e^2}{mc} . Nf,$$

Eq. (47) becomes $k_\nu = \frac{\pi e^2}{mc} . Nf \frac{2\pi \Delta\nu_N}{4\pi^2 (\nu - \nu_0)^2},$

or $(n\kappa) = \frac{\lambda_0}{4\pi} k_\nu = \frac{1}{8} . \frac{1}{2\pi \nu_0} . \frac{4\pi e^2 Nf}{m} . \frac{2\pi \Delta\nu_N}{4\pi^2 (\nu - \nu_0)^2},$

which, in the classical notation, becomes

$$(n\kappa) = \frac{\rho \nu'}{8\omega_0 \mu^2} \qquad \text{......(64)},$$

where $$\rho = \frac{4\pi e^2 Nf}{m} \qquad \text{......(65)},$$

$$\nu' = 2\pi \Delta\nu_N \left(= \frac{1}{\tau} \right) \qquad \text{......(66)},$$

$$\omega_0 = 2\pi \nu_0 \qquad \text{......(67)},$$

$$\mu = 2\pi (\nu - \nu_0) \qquad \text{......(68)}.$$

It is clear from Eq. (64) that if $\rho\nu'$ is determined by measuring $(n\kappa)$ and μ, and if ρ is determined at the same time by some other experiment, ν' can be finally calculated.

Minkowski[49] passed a continuous spectrum through a long column of sodium vapour at various vapour pressures and photographed the D lines in absorption. Plotting the absorption coefficient against the frequency he was able to show that, in the vapour pressure region from 0·0053 mm. to 0·0087 mm., the absorption coefficient obeyed Eq. (64); whereas at lower vapour pressures, the absorption line was too narrow to be resolved properly by the slit of the spectrograph, and at higher vapour pressures Holtsmark coupling broadening made the line so broad that Eq. (64) was invalidated. In the region of vapour pressure in which Eq. (64) is valid, Minkowski measured $n\kappa$ and μ^2 and calculated therefrom $\rho\nu'$ at a number of vapour pressures. From experiments on magneto-rotation, which will be described later, he obtained ρ at these vapour pressures, and combining the results, obtained ν' and consequently τ.

4 j.   TOTAL ENERGY ABSORBED FROM A CONTINUOUS
SPECTRUM BY A RESONANCE LINE THAT IS NOT COMPLETELY
RESOLVED.   If a continuous spectrum is passed through an
absorbing column of gas and the intensity of radiation transmitted is plotted against the frequency in the neighbourhood of
an absorption line, the curve obtained may appear as the heavy
curve in Fig. 26. Ladenburg and Reiche [35] defined as the
"Total Absorption" ("Gesamtabsorption"), A_G, 2π *times the*
ratio of the absorbed energy to the incident intensity. If the in-

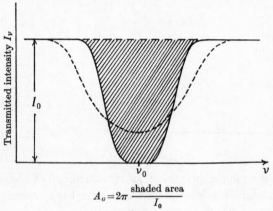

$$A_G = 2\pi \frac{\text{shaded area}}{I_0}$$

Fig. 26. Meaning of "total absorption".

cident intensity is I_0, the absorption coefficient k_ν, and the
thickness of the absorbing column l, the absorbed energy is

$$I_0 \int_0 (1 - e^{-k_\nu l})\, d\nu,$$

whence
$$A_G = 2\pi \int_0^\infty (1 - e^{-k_\nu l})\, d\nu \qquad \ldots\ldots(69),$$

which has the units of 2π times frequency. If the pressure of
the absorbing gas and the thickness of the absorbing layer are
large enough to absorb completely the central region of the
line but not high enough to produce Holtsmark broadening,
then the absorption coefficient is given by Eq. (64), namely

$$k_\nu = \frac{4\pi}{\lambda_0} n\kappa = \frac{\rho\nu'}{4c\mu^2} \qquad \ldots\ldots(70).$$

In order to evaluate Eq. (69), it is convenient to replace the continuous spectrum, of intensity I_0, by a Gauss error curve distribution of intensity $I_0 e^{-q^2 \mu^2}$, where q is a number which, when later allowed to approach zero, will make the distribution of intensity continuous. Since $2\pi d\nu = d\mu$, A_G is then given by

$$A_G = \int_{-\infty}^{\infty} (e^{-q^2 \mu^2} - e^{-q^2 \mu^2 - \frac{\rho\nu' l}{4c\mu^2}}) d\mu$$

$$= \frac{\sqrt{\pi}}{q} [1 - e^{-2q\sqrt{\frac{\rho\nu' l}{4c}}}].$$

As q approaches zero

$$A_G \rightarrow \frac{\sqrt{\pi}}{q} \left[1 - \left(1 - 2q \sqrt{\frac{\rho\nu' l}{4c}} \right) \right],$$

whence, for a continuous spectrum,

$$A_G = \sqrt{\frac{\pi\rho\nu' l}{c}} \qquad \ldots\ldots(71).$$

It must be emphasized that the above equation holds only when ρl (which depends upon the pressure and the thickness of the absorbing layer) is large enough to warrant the use of Eq. (70). The advantage of using Eq. (71) to calculate $\rho\nu'$ is that the measurement of the shaded area in Fig. 26 is, within limits, independent of the width of the slit of the spectrograph. If the absorption line depicted by the heavy curve in Fig. 26 is not completely resolved, the photometer curve may appear as the dotted curve in this figure. Minkowski showed that the area above the dotted curve was, for several values of the slit-width, equal to the shaded area, within the limits of experimental error.

This method of measuring ν' was first employed by Ladenburg and Senftleben[34] in connection with a sodium flame at atmospheric pressure. The result will be discussed in the next chapter under "Lorentz Broadening". Using pure sodium vapour, Minkowski[49] measured the total absorption and obtained with the aid of Eq. (71) the quantity $\rho\nu'$ for the sodium D lines at various vapour pressures, and compared these values with those obtained by the method of Section 4i. The photometer curves obtained by Minkowski in both methods did not show the slightest asymmetry, because the frequency distance from the centre of the D lines at which the measurements were made was so large in comparison to the hyperfine-structure separation.

Since in Eq. (71) ν' is a constant equal to $1/\tau$, and ρ contains Nf, A_G is a convenient measure of Nf, the number of dispersion

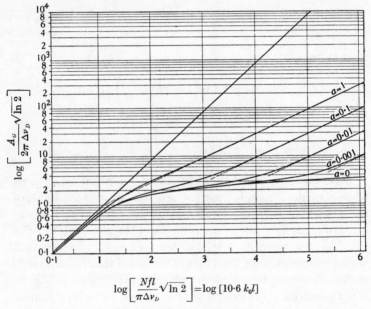

$$\log\left[\frac{Nfl}{\pi\Delta\nu_D}\sqrt{\ln 2}\right] = \log\left[10\cdot6\ k_0 l\right]$$

Fig. 27. Van der Held's theoretical curves of total absorption against number of absorbing atoms.

electrons associated with the absorption and emission of a particular spectral line. The number of dispersion electrons associated with the blue caesium doublet, 4593 and 4555, was measured by Schütz[67], who verified that A_G varies as the square root of Nf at high caesium vapour pressures. If the general expression for the absorption coefficient given by Eq. (40) is substituted in Eq. (69), and A_G calculated by graphical integration for various values of the parameters, the resulting curve enables one to obtain a measure of the number of dispersion electrons corresponding to *any* experimentally measured value of A_G, no matter what the vapour pressure is. This was done by Schütz [68], who, in a summary of the subject in the *Zeitschrift für Astrophysik*, has given a curve of A_G against Nfl for three different values of the natural damping ratio. A similar group of curves was computed by van der Held [22] for four different values of the natural damping ratio. These curves are shown in Fig. 27. The natural damping ratio, denoted by Schütz by the symbol ω'/b and by van der Held by the symbol a, is exactly twice the quantity a appearing in Eq. (40). The values of a shown on the figure are those of van der Held.

A_G was measured by van der Held and Ornstein [23] at a number of values of the sodium vapour pressure, and the experimental curve of A_G against Nfl was compared with van der Held's theoretical curves. The experiments indicated a value of the natural damping ratio equal to 0·005 (within 8 per cent.), corresponding to a lifetime of $1·6 \times 10^{-8}$ sec.

5. MAGNETO-ROTATION AT THE EDGES OF A RESONANCE LINE

5*a*. MAGNETO-ROTATION AT THE EDGES OF A RESOLVED RESONANCE LINE. If a beam of plane polarized light be allowed to traverse longitudinally an absorbing gas placed between the poles of a magnet, the plane of polarization of a particular frequency in the neighbourhood of an absorption line will be rotated by an angle χ_ν. The theory of this phenomenon, which is the familiar Faraday effect, was first worked out by Voigt [76], and later extended by Kuhn [32]. A simple account of Kuhn's theory is given in the Appendix. On the basis of this theory, the angle of rotation at a frequency distance from the centre of the line, which is large in comparison with the separation of the Zeeman components of the line, is given by

$$\chi_\nu = -\frac{\pi e^3 H l z}{2m^2 c^2} \cdot \frac{Nf}{\mu^2} \qquad \ldots \ldots (72),$$

where H is the strength of the magnetic field, l the length of the absorbing column of gas, μ equal to $2\pi (\nu - \nu_0)$, and z is a function of the separations and relative intensities of the Zeeman components. When the magnetic field intensity is in the neighbourhood of 1000 gauss, z is a constant for a particular spectral line. A complete discussion of the quantity z along with a table of values for various resonance lines is given in the Appendix.

Introducing the classical quantity $\rho = \dfrac{4\pi e^2 Nf}{m}$, Eq. (72) becomes

$$\chi_\nu = -\frac{ezH\rho l}{8mc^2\mu^2} \qquad \ldots\ldots(73),$$

which enables one to compute ρ from experimental measurements of χ_ν, μ^2 and H. If N is known, f can be calculated from ρ. If not, the value of ρ at a particular pressure is combined with the value of $\rho\nu'$ at the same pressure obtained by the methods of Sections $4i$ or $4j$, to yield finally a value of ν'. It must be emphasized that Eq. (73) is valid only when μ is large, that is, at great frequency distances from the centre of the line. In order that the angle χ_ν shall have a measurable value at the extreme edges of the absorption line, the pressure of the absorbing gas and the length of the absorbing column (that is, ρl) must be made rather large. These are usually chosen so that the centre of the line is completely absorbed. It is also necessary that the resolving power of the spectroscope or spectrograph must be large enough to enable one to measure the exact value of μ at which a particular angle of rotation occurs. The choice of polarization apparatus depends on the spectral region in which one is working. For details as to the measurement of χ_ν the reader is referred to the papers of Kuhn[32] and Minkowski[49].

The first exact measurements of ρ by the method of magneto-rotation were made by Ladenburg and Minkowski[38], who measured the magneto-rotation at the edges of the sodium D lines at various vapour pressures. Combining the resulting values of ρ with values of the sodium vapour pressure, f was found to be very nearly equal to 1/3 for the D_1 line and 2/3 for the D_2 line, or, what amounts to the same thing, assuming f_{D_1} to be 1/3 and f_{D_2} to be 2/3, the curve of ρ against temperature was shown to be in agreement with the vapour pressure curve.

Minkowski[49] measured the magneto-rotation at the edges of the sodium D lines at various vapour pressures and combined the resulting values of ρ with his own values of $\rho\nu'$ obtained by the methods of Sections $4i$ and $4j$. The result of these three investigations was a value of ν' for both lines equal to 0.62×10^8, yielding a value for τ equal to 1.6×10^{-8} sec. In the same way, Schütz measured ν' for the blue caesium doublet.

From measurements of magneto-rotation and vapour pressure, Minkowski and Mühlenbruch[50] obtained ρ and N for the two caesium resonance lines ($6\,^2S_{1/2}-6\,^2P_{1/2}$, 8944; and $6\,^2S_{1/2}-6\,^2P_{3/2}$, 8521), yielding a value of $f = 0.32$ for the first and 0.66 for the second, corresponding to $\tau = 3.8 \times 10^{-8}$ sec. and 3.3×10^{-8} sec. respectively. In the same way Kuhn[32] obtained for the two cadmium resonance lines ($5\,^1S_0-5\,^1P_1$, 2288; and $5\,^1S_0-5\,^3P_1$, 3261), $f = 1.20$ and 0.0019 respectively, corresponding to $\tau = 2.0 \times 10^{-9}$ sec. and 2.5×10^{-6} sec., and for the two thallium resonance lines ($6\,^2P_{1/2}-6\,^2D_{3/2}$, 2768; and $6\,^2P_{1/2}-7\,^2S_{1/2}$, 3776), $f = 0.20$ and 0.08. Similarly, Weiler[80] found for the two potassium resonance lines ($4\,^2S_{1/2}-4\,^2P_{1/2}$, 7699; and $4\,^2S_{1/2}-4\,^2P_{3/2}$, 7665), $f = 0.33$ and 0.67.

5b. MAGNETO-ROTATION AND ABSORPTION OF A RESONANCE LINE THAT IS NOT COMPLETELY RESOLVED. If a continuous spectrum be sent in turn through a polarizing Nicol, an absorption tube placed longitudinally between the poles of a magnet, an analysing Nicol and a spectroscope which does not resolve the absorption line of the gas in the absorption tube, the whole field of view will be dark when the two Nicols are crossed and when the magnetic field is zero. If the Nicols *are kept crossed*, and the magnetic field is turned on, the amount of light of a particular frequency that passes through the analysing Nicol will depend upon (1) the angle χ_ν through which the plane of polarization of that wave-length has been rotated, and (2) the absorption coefficient k_ν of the gas for that wave-length. If the pressure of the absorbing gas and the length of the absorption tube are great enough to absorb completely the central region of the line, but not great enough to produce Holtsmark broadening, then, from Eq. (73),

$$\chi_\nu = -\frac{ezH\rho l}{8mc^2\mu^2},$$

and, from Eq. (70), $k_\nu = \dfrac{\rho\nu'}{4c\mu^2}.$

If the intensity of the continuous spectrum be I_0, and the length of the absorption tube be l, the total intensity of the light J passing through the analysing Nicol will be

$$J = I_0 \int_{-\infty}^{\infty} \sin^2 \chi_\nu\, e^{-k_\nu l}\, d\mu \qquad \ldots\ldots(74),$$

or $$J = I_0 \int_{-\infty}^{\infty} \sin^2 \frac{ezH\rho l}{8mc^2\mu^2} \cdot e^{-\frac{\rho\nu'l}{4c\mu^2}}\, d\mu \qquad \ldots\ldots(75).$$

Letting $$x^2 = \frac{ezH\rho l}{8mc^2} \cdot \frac{1}{\mu^2},$$

$$b = \frac{2mc\nu'}{ezH},$$

Eq. (75) becomes

$$J = I_0 \sqrt{\frac{ezH\rho l}{8mc^2}} \int_{-\infty}^{\infty} \frac{1}{x^2} \sin^2 x^2\, e^{-bx^2}\, dx \qquad \ldots\ldots(76),$$

and since

$$\int_{-\infty}^{\infty} \frac{1}{x^2} \sin^2 x^2\, e^{-bx^2}\, dx = \frac{\sqrt{2\pi b}}{4} \left[\sqrt{1 + \sqrt{1 + \frac{4}{b^2}}} - \sqrt{2} \right] \ldots(77),$$

the final result is

$$\frac{J}{I_0} = \frac{1}{4} \sqrt{\frac{\pi\rho\nu'l}{2c}} \left[\sqrt{1 + \sqrt{1 + \left(\frac{ezH}{mc\nu'}\right)^2}} - \sqrt{2} \right] \quad \ldots(78).$$

It is apparent from the above equation that a measurement of J/I_0 and H, along with either a measurement or a calculation of ρ, is sufficient to enable one to calculate ν'. This method was employed by Schütz [66] to measure ν' for the two sodium D lines. He extended Eq. (78) to include both lines which were not separated by his spectroscope, and calculated ρ from Minkowski's measurements of f and from the vapour pressure. He obtained the result that $\nu' = 0.64 \times 10^8$ ($\tau = 1.6 \times 10^{-8}$ sec.) in the sodium vapour pressure interval from 6.6×10^{-4} mm. to

$3 \cdot 7 \times 10^{-3}$ mm., and that ν' increased at higher vapour pressure because of Holtsmark broadening. Further results of Schütz in connection with collision broadening by foreign gases will be discussed later on in this book.

The disadvantage of Schütz's method of determining ν' is the necessity for knowing ρ. An extremely ingenious variation of Schütz's method was developed by Weingeroff[81], in which a knowledge of ρ is not necessary. It will be remembered that the method of Schütz involved the measurement of the total intensity of light transmitted through the analysing Nicol when both Nicols were crossed. Weingeroff noticed that, when the magnetic field was at some constant value and the analysing Nicol was rotated in the direction of the magneto-rotation, the observed line was first bright on a dark background, then it vanished into the background, and then it appeared dark on a bright background. A similar series of changes occurred when the analysing Nicol was turned in a direction opposite to the magneto-rotation. If the angle ϕ denote the position of the analysing Nicol with reference to the crossed position (when Nicols are crossed $\phi = 0$), then the difference between the amount of light due to magneto-rotation plus background and the amount due to background alone is given by

$$R = I_0 \int_{-\infty}^{\infty} [\sin^2 (\chi_\nu + \phi) e^{-k_\nu l} - \sin^2 \phi] \, d\mu \ \ \ldots\ldots (79),$$

which obviously reduces to Schütz's expression, Eq. (74), when $\phi = 0$. When the central region of the line is completely absorbed, χ_ν and k_ν can be expressed as before, and Eq. (79) becomes

$$R = I_0 \int_{-\infty}^{\infty} \left[\sin^2 \left(\frac{ezH\rho l}{8mc^2\mu^2} + \phi \right) e^{-\frac{\rho\nu' l}{4c\mu^2}} - \sin^2 \phi \right] d\mu \ \ldots (80),$$

which, upon substituting

$$x^2 = \frac{ezH\rho l}{8mc^2} \cdot \frac{1}{\mu^2},$$

$$b = \frac{2mc\nu'}{ezH},$$

reduces to

$$R = I_0 \sqrt{\frac{ezH\rho l}{8mc^2}} \int_{-\infty}^{\infty} \left[\frac{1}{x^2} \sin^2 (x^2 + \phi) e^{-bx^2} - \sin^2 \phi \right] dx \ldots (81).$$

Since

$$\int_{-\infty}^{\infty} \left[\frac{1}{x^2} \sin^2 (x^2 + \phi) \, e^{-bx^2} - \sin^2 \phi \right] dx$$

$$= \frac{\sqrt{2\pi b}}{4} \left\{ \cos 2\phi \sqrt{1 + \sqrt{1 + \frac{4}{b^2}}} + \sin 2\phi \sqrt{-1 + \sqrt{1 + \frac{4}{b^2}}} - \sqrt{2} \right\}$$

$$\dots\dots(82),$$

$$R = \frac{I_0}{4} \sqrt{\frac{\pi l \rho \nu'}{2c}} \left\{ \cos 2\phi \sqrt{1 + \sqrt{1 + \frac{4}{b^2}}} \right.$$

$$\left. + \sin 2\phi \sqrt{-1 + \sqrt{1 + \frac{4}{b^2}}} - \sqrt{2} \right\} \dots\dots(83).$$

The above equation gives the amount of light over and above the background that passes through the analysing Nicol when the magnetic field is H and the setting of the analysing Nicol is ϕ. If H is kept constant and ϕ varied, there will be a value of ϕ for which R will vanish, that is, the line will merge with the background. Let this value of ϕ be denoted by ϕ_0. Then ϕ_0 is given by the equation

$$\cos 2\phi_0 \sqrt{1 + \sqrt{1 + \frac{4}{b^2}}} + \sin 2\phi_0 \sqrt{-1 + \sqrt{1 + \frac{4}{b^2}}} - \sqrt{2} = 0$$

$$\dots\dots(84),$$

whose solution was obtained graphically by Weingeroff and is shown in Fig. 28 in which ϕ_0 is plotted as abscissa and $b = \frac{2mc\nu'}{ezH}$ plotted as ordinate. The experiment consists in measuring the angle ϕ_0 through which the analysing Nicol must be rotated in order that the line merge with the background. From Fig. 28 the corresponding value of $\frac{2mc\nu'}{ezH}$ is read off, whence, knowing z and H, ν' is calculated. The great advantage of this method lies in the fact that a knowledge of the vapour pressure (that is, ρ) is not necessary. In this way Weingeroff measured ν' for the sodium D lines at various vapour pressures. Since the two D lines were not separated by his spectroscope, Weingeroff extended Eq. (83) to include both lines. Furthermore, since it was necessary to work at low magnetic field intensities where

z is not constant but is a function of H, it was necessary to perform other experiments to obtain z. The final result was that in the vapour pressure range corresponding to temperatures from 240° C. to 330° C., ν' remained constant at 0.62×10^8 ($\tau = 1.6 \times 10^{-8}$ sec.), and beyond 330° C. it increased because of Holtsmark broadening. Schütz found that ν' began to increase

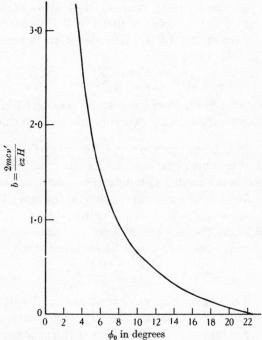

Fig. 28. Graph of
$$\cos 2\phi_0 \sqrt{1 + \sqrt{1 + \frac{4}{b^2}}} + \sin 2\phi_0 \sqrt{-1 + \sqrt{1 + \frac{4}{b^2}}} - \sqrt{2} = 0.$$

at a vapour pressure corresponding to a temperature of 265° C., and Minkowski at a temperature of 287° C. Of these three results, that of Weingeroff is probably the most accurate.

6. DISPERSION AT THE EDGES OF A RESONANCE LINE

6*a*. GENERAL DISPERSION FORMULA. On the basis of Kramers' quantum-theoretical dispersion formula, Ladenburg [39] showed that the index of refraction of a gas at a wave-length λ is given by

$$n - 1 = \frac{e^2}{2\pi mc^2} \sum_{j=0}^{\infty} \sum_{k=j+1}^{\infty} \frac{\lambda^2 \lambda^2_{kj}}{\lambda^2 - \lambda^2_{kj}} N_j f_{kj} \left(1 - \frac{N_k}{N_j} \frac{g_j}{g_k}\right) \quad \ldots\ldots(85),$$

where k and j refer to any two stationary states (k being the upper state) whose statistical weights are g_k and g_j respectively. N_k and N_j are the numbers of atoms in the two states, λ_{kj} the wave-length of the radiation emitted in the transition $k \to j$, and f_{kj} is connected with the Einstein A coefficient by the formula [see Eq. (31)]

$$f_{kj} = \frac{mc\lambda^2_{kj}}{8\pi^2 e^2} \cdot \frac{g_k}{g_j} A_{k \to j}.$$

It is convenient to consider three special cases of Eq. (85), in order to discuss the existing experimental work in this field.

6b. NORMAL DISPERSION OF AN UNEXCITED GAS VERY FAR FROM THE ABSORPTION LINES. If the gas is not electrically excited and is at a moderate temperature, there will be only a negligible number of atoms in excited states other than the normal one. Denoting the normal state by $j = 1$, and calling $N_j = N_1 = N$, Eq. (85) can be simplified by neglecting the ratio N_k/N_j in comparison with unity. There results then

$$n - 1 = \frac{e^2 N}{2\pi mc^2} \sum_{k=2}^{\infty} \frac{\lambda^2 \lambda^2_{k1} f_{k1}}{\lambda^2 - \lambda^2_{k1}} \quad \ldots\ldots(86),$$

where λ_{k1} (i.e. λ_{21}, λ_{31}, λ_{41}, etc.) are the wave-lengths of the absorption lines that influence the dispersion, and f_{k1} are the respective f values of these lines. This is the ordinary normal dispersion curve first derived classically by Sellmeyer, and gives the value of the index of refraction at wave-lengths that are hundreds or thousands of Ångströms away from the absorption lines. In this region, n is most easily measured by the method of Puccianti [59], involving the use of a Jamin interferometer.

It was shown by Herzfeld and Wolf [24] that the existing values of n for each inert gas He, Ne, A, Kr and Xe *in the visible region* could be represented by an equation of the type

$$n - 1 = \frac{\text{const.}}{\nu_0^2 - \nu^2}.$$

The values of λ_0 calculated from the empirical curves, however, did not agree with the known ultra-violet resonance lines of the noble gases, and were, in fact, in all cases, of shorter wavelength than the series limit. Upon attempting to use a dispersion formula of two terms, one term involving the correct ultra-violet resonance frequency, it was found that the wavelength of the second absorption line did not correspond to any known absorption line, being also of much too short a wavelength. It is therefore apparent that measurements of index of refraction at wave-lengths that are too far removed from the wave-lengths at which the absorption lines occur are not very reliable in giving information concerning the absorption lines themselves.

A very careful measurement of the index of refraction of mercury vapour was made by Wolfsohn[85] at wave-lengths from 2700 to 7000. He found that the results could be represented by Eq. (86), using three terms, the first two involving the two ultra-violet resonance lines of mercury, 1850 and 2537, and the third involving a wave-length somewhere between 1400 and 1100. Inserting in the formula the accurate f-value for 2537 obtained by measurements of anomalous dispersion (to be described later), he found that the f-value for 1850 varied from 0·7 to 1·0 depending upon the wave-length chosen for the third term. Choosing the third absorption line to be at 1190, the limit of the principal series of mercury, he obtained [43] for the f-value of the 1850 line 0·96 corresponding to a lifetime of the $6\,^1P_1$ state of $1·6 \times 10^{-9}$ sec.

6c. Anomalous Dispersion of an Unexcited Gas at the Edges of a Resonance Line. If, instead of measuring n at wave-lengths that are hundreds or thousands of Ångströms from an absorption line, the index is measured from 0·5 to 1 Ångström from the centre of the resonance line at λ_{21}, the effect of all the other absorption lines becomes negligible, and Eq. (86) may be further simplified by not having to sum over k. Denoting λ_{21} by λ_0 and f_{21} by f, Eq. (86) becomes

$$n - 1 = \frac{e^2 N f}{2\pi m c^2} \cdot \frac{\lambda^2 \lambda_0^2}{\lambda^2 - \lambda_0^2},$$

or
$$n - 1 = \frac{e^2 Nf}{4\pi mc^2} \cdot \frac{\lambda_0{}^3}{\lambda - \lambda_0} \qquad \ldots\ldots(87).$$

This is the well-known formula of anomalous dispersion, and has been used in conjunction with experiment to provide some of the best f-values that have as yet been obtained. The most accurate experimental method is the "hook-method" of Roschdestwensky[61, 62], which will be described in detail. In Fig. 29 is shown a schematic diagram of a Jamin interferometer.

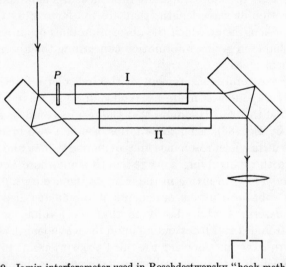

Fig. 29. Jamin interferometer used in Roschdestwensky "hook-method".

Gas may be admitted into tube I at any known pressure, but tube II is kept evacuated. A source of continuous radiation is used, and the resulting beam of light is focused on the slit of a spectroscope. With both tubes evacuated, and with the compensating plate P removed, the continuous spectrum is crossed by horizontal interference fringes. With the compensating plate in position, the interference fringes are oblique. If the wave-length separation of a convenient number of fringes in the immediate neighbourhood of λ_0 is measured, an important constant of the apparatus K can be calculated as follows:

$$K = -\lambda_0 \frac{\text{No. of fringes}}{\text{Wave-length separation of these fringes}}.$$

If a gas with an absorption line at λ_0 is now introduced into tube I, the oblique interference fringes become hook-shaped symmetrically on both sides of the absorption line. If A represent the wave-length separation of two hooks symmetrically placed with regard to the absorption line, then the theory of this method in conjunction with Eq. (87) yields the equation

$$f = \frac{\pi m c^2}{e^2 \lambda_0^3 N l} \cdot K A^2 \qquad \ldots\ldots(88),$$

from which f may be calculated when N and l (the thickness of the layer of gas) have been determined. A simple derivation of Eq. (88) may be found in a paper by Ladenburg and Wolfsohn [41].

Roschdestwensky [62] used the hook-method to study the anomalous dispersion in the neighbourhood of the absorption lines of Na, K, Rb and Cs. The measurements led in all cases to the ratio of the f-values associated with the principal series of doublets. The results are tabulated in Table XII.

TABLE XII

Element	Running numbers n, m	$\dfrac{f \text{ of } n\,^2S_{1/2}-m\,^2P_{3/2}}{f \text{ of } n\,^2S_{1/2}-m\,^2P_{1/2}}$
Na	3, 3	1·98
K	4, 4	1·98
,,	4, 5	2·05
Rb	5, 5	2·01
,,	5, 6	2·57
,,	5, 7	2·9
Cs	6, 6	2·05
,,	6, 7	4·07
,,	6, 8	7·4
,,	6, 9	9·1

By the same method, Ladenburg and Wolfsohn [41] measured the anomalous dispersion of mercury vapour near the resonance line 2537 within a wide range of vapour pressures. Using Eq. (88) for low values of the vapour pressure, and the results of Wolfsohn's investigation of the normal dispersion of mercury vapour at high vapour pressures, the authors obtained a value of f equal to 0.0255 ± 0.005, corresponding to a lifetime of the $6\,^3P_1$ state of 1.14×10^{-7} sec. With the Jamin interferonometer enclosed in an evacuated vessel, Wolfsohn[85 a] measured the anomalous dispersion of mercury vapour in the neighbourhood of the 1849 line ($6\,^1S_0$–$6\,^1P_1$) and obtained for f, 1·19, corresponding to a lifetime of the $6\,^1P_1$ state of $1\cdot30 \times 10^{-9}$ sec.

By means of the Roschdestwensky hook-method, Prokofjew[58] studied the anomalous dispersion of the resonance lines of Ca, Sr and Ba under experimental conditions which did not allow the accurate measurement of vapour pressure. Calling τ_1 the lifetime of the singlet state and τ_2 the lifetime of the triplet state, Prokofjew's experiments yielded the ratio τ_1/τ_2. For Ca, Sr and Ba, this ratio was found to be $1\cdot25 \times 10^{-5}$, $26\cdot9 \times 10^{-5}$ and 335×10^{-5} respectively. In the same way, Filippov[8] found the ratio τ_1/τ_2 of Zn and Cd to be $6\cdot76 \times 10^{-5}$ and $72\cdot5 \times 10^{-5}$ respectively. Prokofjew and Solowjew[57] found the ratio of the f-values of the thallium lines, 5350 and 3776, to be 0·95.

6d. ANOMALOUS DISPERSION OF A STRONGLY EXCITED. GAS AT THE EDGES OF THE ABSORPTION LINES λ_{kj}. If n is measured very close to an absorption line, the effect of the other absorption lines may be neglected, but when the excitation is strong enough, the number of atoms in the higher state N_k may become an appreciable fraction of the number in the lower state N_j. Eq. (85) then becomes

$$n-1 = \frac{e^2}{4\pi mc^2} \frac{\lambda^3_{kj}}{\lambda-\lambda_{kj}} N_j f_{kj} \left(1 - \frac{N_k}{N_j} \cdot \frac{g_j}{g_k}\right) \quad \ldots\ldots(89).$$

The expression $\left(1 - \dfrac{N_k}{N_j} \cdot \dfrac{g_j}{g_k}\right)$ is known as the negative dispersion term and is appreciably different from unity only when the excitation of the gas is very strong. By the hook-method of Roschdestwensky, Ladenburg, Kopfermann and Levy investigated the anomalous dispersion of electrically excited neon, in the neighbourhood of many absorption lines originating at the metastable $^3P_2$ and $^3P_0$ levels. The results of these investigations, in conjunction with the results of intensity measurements of the neon lines, enabled the authors to give relative f-values of all the neon lines studied. It was found that when the current through the discharge tube was greater than 200 milliamps. the negative dispersion term began to play an important rôle. Similar measurements were made on hydrogen by Ladenburg and Carst, and on helium by Levy. For further details the reader is referred to the original papers in volumes **48**, **65** and **72** of the *Zeitschrift für Physik*.

7. TABLES OF LIFETIMES
AND DISCUSSION

7*a*. SUMMARY OF METHODS OF MEASURING LIFETIME, AND TABLES OF LIFETIMES.

Methods involving the emission of radiation.

Decay of electrically excited resonance radiation, § 3*c*.
Decay of optically excited resonance radiation, § 3*d*.
Decay along an optically excited atomic ray, § 3*e*.
Decay along a canal ray, § 3*f*.
Absolute intensity of a resonance line, § 3*g*.

Methods involving absorption of radiation.

Total area of absorption coefficient, § 4*a*.
Absorption coefficient at the centre of a resonance line, § 4*b, c, d*.
Absorption coefficient at the edges of a resonance line, § 4*i*.
Total energy absorbed from a continuous spectrum, § 4*j*.

Methods involving magneto-rotation of polarized light.

Magneto-rotation at the edges of a resolved resonance line, § 5*a*.
Magneto-rotation and absorption of an unresolved resonance line, § 5*b*.

Methods involving dispersion of radiation.

Normal dispersion very far from a resonance line, § 6*b*.
Anomalous dispersion at the edges of a resonance line, § 6*c*.

Methods involving depolarization (see Chap. v).

Depolarization of resonance radiation by a steady magnetic field.
Measurement of the angle of maximum polarization in small steady magnetic fields.

Depolarization of resonance radiation by an alternating magnetic field.

All the results that have appeared throughout the chapter are collected in Tables XIII, XIV and XV.

TABLE XIII

Atom	Series notation of resonance line	Wave-length	τ in secs.	f-value	Author, reference and method
H	$1\,^2S_{1/2}-2\,^2P$	1216	$1\cdot2\ \times10^{-8}$	—	Slack [69], § 3c
He	$1\,^1S_0\,-2\,^1P_1$	584	$4\cdot42\times10^{-10}$	0·349	Vinti [77a], Theor.
,,	,,	,,	$5\cdot80\times10^{-10}$	0·266	Wheeler [83a], Theor.
Li	$2\,^2S_{1/2}-2\,^2P$	6708	$2\cdot7\ \times10^{-8}$	0·25 and 0·50	Trumpy [74a], Theor.
Na	$3\,^2S_{1/2}-3\,^2P$	5896, 5890	$1\cdot5\ \times10^{-8}$	—	Hupfield [27], § 3d
,,	,,	,,	$8\cdot2\ \times10^{-9}$	—	Duschinsky [5], § 3d
,,	,,	,,	$1\cdot6\ \times10^{-8}$	—	v. d. Held and Ornstein [23], § 3g, § 4j
,,	,,	,,	$1\cdot48\times10^{-8}*$	0·35 and 0·70	Ladenburg and Thiele [43a], § 7b
,,	,,	,,	$1\cdot6\ \times10^{-8}$	0·33 and 0·67	Minkowski [49], §5a, 4i
,,	,,	,,	$1\cdot6\ \times10^{-8}$	0·33 and 0·67	Minkowski [49], §5a, 4j
,,	,,	,,	$1\cdot6\ \times10^{-8}$	0·33 and 0·67	Schütz [66], § 5b
,,	,,	,,	$1\cdot6\ \times10^{-8}$	0·33 and 0·67	Weingeroff [81], § 5b
,,	,,	,,	$1\cdot6\ \times10^{-8}$	—	Sugiura [72], Theor.
K	$4\,^2S_{1/2}-4\,^2P$	7699, 7665	$2\cdot7\ \times10^{-8}*$	0·33 and 0·67	Weiler [80], § 5a
Cs	$6\,^2S_{1/2}-6\,^2P_{1/2}$	8944	$3\cdot8\ \times10^{-8}*$	0·32	Minkowski and Mühlenbruch [59], § 5a
,,	$6\,^2S_{1/2}-6\,^2P_{3/2}$	8521	$3\cdot3\ \times10^{-8}*$	0·66	,,

7b. DISCUSSION OF TABLES. The values of f and τ listed in the tables represent for the most part work done in the last eight years. Previous work was concerned mainly with the ratio of the f-values of the principal series doublets of the alkalis. A convenient summary of such work can be found in the *Zürich Habilitationsschrift* of W. Kuhn and in a paper by J. Weiler, *Z. f. Phys.* **50**, 436 (1928). Further information can be found in the articles of R. Minkowski and R. Ladenburg in the section of Miller-Pouillet's *Handbuch der Physik* devoted to optics, in the article on dispersion and absorption by G. Jaffé in the *Handbuch der Experimentalphysik*, Vol. **19**, and

TABLE XIV

Atom	Series notation of resonance line	Wave-length	τ in secs.	f-value	Author, reference and method
Mg	$3\,^1S_0$–$3\,^3P_1$	4571	$\sim 4 \times 10^{-3}$	—	Frayne [9], § 7b
Zn	$4\,^1S_0$–$4\,^3P_1$	3076	$\sim 1 \times 10^{-5}$	—	Soleillet, chap. v
,,	,,	,,	$\sim 1 \times 10^{-5}$	—	Soleillet [71], § 3e
,,	$4\,^1S_0$–$4\,^1P_1$	2139	$< 10^{-7}$	—	Soleillet [71], § 3e
Cd	$5\,^1S_0$–$5\,^3P_1$	3261	$2.5 \times 10^{-6}*$	—	Koenig and Ellett [30], § 3e
,,	,,	,,	$2.5 \times 10^{-6}*$	0.0019	Kuhn [32], § 5a
,,	,,	,,	$\sim 2 \times 10^{-6}$	—	Soleillet, chap. v
,,	,,	,,	$2.3 \times 10^{-6}*$	—	Ellett, chap. v
,,	$5\,^1S_0$–$5\,^1P_1$	2288	$1.98 \times 10^{-9}*$	1.20	Kuhn [32], § 5a
,,	,,	,,	$1.99 \times 10^{-9}*$	—	Zemansky [90], § 4e
,,	,,	,,	$\sim 10^{-9}$	—	Soleillet, chap. v
Tl	$6\,^2P_{3/2}$–$7\,^2S_{1/2}$	5350	$\left\{ \begin{array}{l} \tau \text{ of } 7\,^2S_{1/2} \\ \text{state is} \\ 1.4 \times 10^{-8} \end{array} \right.$	0.076*	Prokofjew and Solowjew [57], § 6c
,,	$6\,^2P_{1/2}$–$7\,^2S_{1/2}$	3776		0.08*	Kuhn [32], § 5a
,,	$6\,^2P_{1/2}$–$6\,^2D_{3/2}$	2768	—	0.20*	Kuhn [32], § 5a

TABLE XV

Atom	Series notation of resonance line	Wave-length	τ in secs.	f-value	Author, reference and method
Hg	$6\,^1S_0$–$6\,^3P_1$	2537	$\sim 1 \times 10^{-7}$	—	Webb and Messenger [79], § 3c
,,	,,	,,	$1.08 \times 10^{-7}*$	—	Garrett [12], § 3c
,,	,,	,,	0.98×10^{-7}	—	Wien [84], § 3f
,,	,,	,,	1.0×10^{-7}	—	Füchtbauer, Joos and Dinkelacker, as calculated by Tolman [11, 74], § 4a
,,	,,	,,	$1.08 \times 10^{-7}*$	0.0278	Kopfermann and Tietze, as calculated by Zehden and Zemansky [31, 87], § 4f
,,	,,	,,	$1.14 \times 10^{-7}*$	0.0255	Ladenburg and Wolfsohn [41], § 6c
,,	,,	,,	1.13×10^{-7}	—	von Keussler, Chap. v
,,	,,	,,	$1.08 \times 10^{-7}*$	—	Olson, recalculated by Mitchell, Chap. v
,,	,,	,,	$\sim 10^{-7}$	—	Breit and Ellett, Chap. v
,	,,	,,	$\sim 10^{-7}$	—	Fermi and Rasetti, Chap. v
Hg	$6\,^1S_0$–$6\,^1P_1$	1849	0.3×10^{-9}	—	Garrett [12], § 4f
,,	,,	,,	1.6×10^{-9}	0.96	Ladenburg and Wolfsohn [43], § 6b
,,	,,	,,	$1.30 \times 10^{-9}*$	1.19	Wolfsohn [85a], § 6c

in an article by Korff and Breit in the *Reviews of Modern Physics*, Vol. 4, No. 3.

The values in the tables which, in the opinion of the authors, are the most accurate are starred. The value for the lifetime of the $3\,^2P$ states of sodium given by Ladenburg and Thiele, 1.48×10^{-8} sec., is suggested as the best value to date, being the result of a critical survey of previous experiments. In the case of the lifetime of the $6\,^3P_1$ state of mercury, the value of 1.08×10^{-7} sec. (Garrett and Webb) is suggested as the most reliable, partly because it is independent of a knowledge of the mercury vapour pressure and also because it agrees best with the absorption measurements of Kopfermann and Tietze when the Schüler and Keyston hyperfine structure is taken into account.

The most accurate value of lifetime is probably that of the $5\,^3P_1$ state of the cadmium atom, 2.5×10^{-6} sec., inasmuch as precisely this value is obtained by two utterly different methods (Kuhn, Koenig and Ellett), and very nearly this value, 2.3×10^{-6} sec. (Ellett), by still a third method. The agreement between Kuhn's and Zemansky's values for the lifetime of the $5\,^1P_1$ state of the cadmium atom suggests that the value 2.0×10^{-9} sec. for this lifetime is quite reliable.

The lifetime of the $7\,^2S_{1/2}$ state of thallium was obtained as follows. From Kuhn's measurement of the f-value of the 3776 line, and from Ladenburg's equation connecting the f-value with the Einstein A coefficient, the result was obtained:

$$A_{3776} = 3.7 \times 10^7.$$

Vonwiller[77] measured the ratio of the intensity of the 3776 line to that of the 5350 line and obtained the value 1.56. From the relation

$$\frac{I_{3776}}{I_{5350}} = \frac{A_{3776}}{A_{5350}} \cdot \frac{\nu_{3776}}{\nu_{5350}} = 1.56,$$

A_{5350} was found to be 3.4×10^7. Prokofjew and Solowjew[57] measured the ratio of the f-values of the 5350 and 3776 lines and obtained the value 0.95. From the relation

$$\frac{f_{5350}}{f_{3776}} = \frac{A_{5350}}{A_{3776}} \cdot 2 \cdot \left(\frac{5.35}{3.78}\right)^2 = 0.95,$$

A_{5350} was found to be $3 \cdot 5 \times 10^7$, in good agreement with the first value. The sum of the two Einstein A's is

$$3 \cdot 7 \times 10^7 + 3 \cdot 5 \times 10^7 = 7 \cdot 2 \times 10^7,$$

and since $\tau = 1/\Sigma A$, the lifetime of the $7\,^2S_{1/2}$ state is $1 \cdot 4 \times 10^{-8}$ sec.

The value of the lifetime of the $3\,^3P_1$ level of magnesium calculated by Frayne[9] on the basis of Houston's wave-mechanical formulas is given in the table, because it is partly substantiated by Frayne's experiments on the emission characteristics of a magnesium arc. These experiments indicate that the intensity of the 4571 line emitted by magnesium vapour in the presence of foreign gases is consistent with a value of τ equal to 4×10^{-3} sec.

7c. ELECTRON EXCITATION FUNCTIONS. A possible connection between lifetime and electron excitation function was found by Hanle[21], Schaffernicht[64a] and Larché[44]. If the intensity of a spectral line be plotted against the electron excitation voltage a curve is obtained which starts in the neighbourhood of the excitation potential, rises to a maximum and then decreases to zero as the voltage is increased. Such a curve is called an excitation curve, and in the case of a resonance line there seems to be a relation between the width of the maximum and the lifetime. The excitation curves for the intercombination lines of Hg, Cd and Zn rise quickly and descend quickly with a width that is quite narrow. The curves for the singlet resonance lines, however, descend very slowly, making the width rather large. The authors conclude that short lifetimes are associated with wide excitation curves, and long lifetimes with narrow excitation curves. More experimental material, however, is needed to express this regularity in a more quantitative form. The excitation curve of the cadmium line 2288 $(5\,^1S_0 - 5\,^1P_1)$ obtained by Larché showed an interesting anomaly. It has a slight depression as though it were the sum of two curves, one thin and the other thick. On the basis of the relation just expressed between lifetime and width of excitation curve, this seems to indicate that the $5\,^1P_1$ level of cadmium has two lifetimes, of the order of 10^{-9} sec. and 10^{-6} sec.

respectively, which is exactly what was inferred by Soleillet[70] from his experimental curves expressing the percentage polarization of cadmium resonance radiation 2288 and external magnetic field. There seems to be no doubt about the experimental results of Larché and Soleillet, but there is considerable objection to supposing that the hyperfine structure levels of the $5\,{}^1P_1$ state of cadmium, which are known to be extraordinarily close together, should have lifetimes differing by a factor of 1000. No other explanation, however, has as yet been given.

In contradiction to the experimental results of Hanle, Larché and others, Michels[46] obtained narrow excitation curves for all lines, whether singlet or intercombination lines. It is therefore a possibility that the apparent relation between lifetime and width of maximum is illusory. It is impossible to decide the question at this time.

7 d. THE PAULI-HOUSTON FORMULA. An interesting relation was derived by Pauli on the basis of the correspondence principle and later by Houston on the basis of the wave mechanics.

If $\tau_1 =$ lifetime of the first ${}^1P_1$ state of a 2 electron atom,

$\tau_2 =$ lifetime of the first ${}^3P_1$ state of a 2 electron atom,

$\Delta\nu =$ frequency separation of ${}^3P_0$ and ${}^3P_2$,

$\delta\nu =$ frequency separation of ${}^1P_1$ and centre of gravity of ${}^3P$ states (one-third the way from ${}^3P_0$ to ${}^3P_2$),

the Pauli-Houston formula states that

$$\frac{\tau_1}{\tau_2} = \frac{A_2}{A_1} = \frac{2}{9}\left(\frac{\Delta\nu}{\delta\nu}\right)^2\left(\frac{\nu_2}{\nu_1}\right)^3 \qquad \ldots\ldots(90).$$

In Table XVI, columns (3) and (5), values of τ_1/τ_2 calculated by Eq. (90) are compared with the experimental measurements of this ratio. It is seen that in all cases the calculated values are larger than the experimental ones. In a private communication it was pointed out by Houston that the experimental values of τ_1/τ_2 are in better agreement with a "fourth power of the frequency" law than with Eq. (90). Values of the expression

$$\frac{2}{9}\left(\frac{\Delta\nu}{\delta\nu}\right)^2\left(\frac{\nu_2}{\nu_1}\right)^4$$

are listed in column (4) of Table XVI, and are seen to be in fairly good agreement with measured values of τ_1/τ_2. No explanation, however, has as yet been given for this agreement.

TABLE XVI

Atom	(1) $\left(\dfrac{\Delta\nu}{\delta\nu}\right)$	(2) $\dfrac{\nu_2}{\nu_1}$	(3) $\dfrac{2}{9}\left(\dfrac{\Delta\nu}{\delta\nu}\right)^2\left(\dfrac{\nu_2}{\nu_1}\right)^3$	(4) $\dfrac{2}{9}\left(\dfrac{\Delta\nu}{\delta\nu}\right)^2\left(\dfrac{\nu_2}{\nu_1}\right)^4$	(5) $\dfrac{\tau_1}{\tau_2}=\dfrac{A_2}{A_1}$ measured	Authors
Mg	$4\cdot63\times10^{-3}$	·624	$11\cdot6\times10^{-7}$	$7\cdot24\times10^{-7}$		
Ca	$1\cdot87\times10^{-2}$	·643	$2\cdot06\times10^{-5}$	$1\cdot33\times10^{-5}$	$1\cdot25\times10^{-5}$	Prokofjew [58]
Zn	$4\cdot07\times10^{-2}$	·695	$12\cdot4\times10^{-5}$	$8\cdot64\times10^{-5}$	$6\cdot71\times10^{-5}$	Filippov [8]
Sr	$8\cdot08\times10^{-2}$	·668	$4\cdot31\times10^{-4}$	$2\cdot88\times10^{-4}$	$2\cdot69\times10^{-4}$	Prokofjew [58]
Cd	$1\cdot32\times10^{-1}$	·702	$13\cdot4\times10^{-4}$	$9\cdot40\times10^{-4}$	$8\cdot0\ \times10^{-4}$	Kuhn [32]
					$7\cdot25\times10^{-4}$	Filippov [8]
Ba	$2\cdot32\times10^{-1}$	·700	$4\cdot12\times10^{-3}$	$2\cdot88\times10^{-3}$	$3\cdot35\times10^{-3}$	Prokofjew [58]
Hg	$4\cdot48\times10^{-1}$	·728	$1\cdot73\times10^{-2}$	$1\cdot26\times10^{-2}$	$1\cdot2\ \times10^{-2}$	Garrett [12]
						Wolfsohn [85a]

7 e. HIGHER SERIES MEMBERS OF THE ALKALIS. In working with the resonance radiation of the alkali vapours it is sometimes necessary to know the absorption or emission of the next doublet in the same series. In all cases the f-value and the transition probability of the next doublet are much smaller than those of the resonance lines. This is shown in Table XVII.

TABLE XVII

Atom	Atomic number	$\dfrac{f_{2\to1}}{f_{3\to1}}$	$\dfrac{A_{2\to1}}{A_{3\to1}}$	Author and reference
Li	3	136·5	31·4	A. Filippov [7]
Na	11	69·5	21·8	A. Filippov and W. Prokofjew [6]
K	19	111·5	30·3	W. Prokofjew and G. Gamow [56]
Rb	37	70·3	20·3	D. Roschdestwensky [62]
Cs	55	69·0	19·2	R. Minkowski and W. Mühlenbruch [50]

REFERENCES TO CHAPTER III

[1] Abraham, H. and Lemoine, J., *Compt. Rend.* **129**, 206 (1899).
[2] Compton, K. T., *Phys. Rev.* **20**, 283 (1922).
[3] —— *Phil. Mag.* **45**, 752 (1923).
[4] Dunoyer, L., *Le Rad.* **10**, 400 (1913).
[5] Duschinsky, F., *Z. f. Phys.* **78**, 586 (1932).
[5a] —— *ibid.* **81**, 7, 23 (1933).

[6] Filippov, A. and Prokofjew, W., *Z. f. Phys.* **56**, 458 (1929).

[7] Filippov, A., *ibid.* **69**, 526 (1931).

[8] —— *Sow. Phys.* **1**, 289 (1932).

[9] Frayne, J. G., *Phys. Rev.* **34**, 590 (1929).

[10] Füchtbauer, C., *Phys. Zeits.* **21**, 322 (1920).

[11] Füchtbauer, C., Joos, G. and Dinkelacker, O., *Ann. d. Phys.* **71**, 204 (1923).

[12] Garrett, P. H., *Phys. Rev.* **40**, 779 (1932).

[13] Gaviola, E., *Ann. d. Phys.* **81**, 681 (1926).

[14] —— *Z. f. Phys.* **42**, 853 (1927).

[15] Goos, F. and Meyer, H., *ibid.* **35**, 803 (1926).

[16] Gouy, G. L., *Ann. Chim. Phys.* **18**, 5 (1879).

[17] —— *Compt. Rend.* **88**, 420 (1879).

[18] —— *ibid.* **154**, 1764 (1912).

[19] de Groot, W., *Physica*, **9**, 263 (1929).

[20] v. Hamos, L., *Z. f. Phys.* **74**, 379 (1932).

[21] Hanle, W., *ibid.* **56**, 94 (1929).

[22] v. d. Held, E. F. M., *ibid.* **70**, 508 (1931).

[23] v. d. Held, E. F. M. and Ornstein, S., *ibid.* **77**, 459 (1932).

[24] Herzfeld and Wolf, *Handb. d. Experimentalphysik*, **19**, 89.

[25] Hoyt, F. C., *Phys. Rev.* **36**, 860 (1930).

[26] Hughes, A. L. and Thomas, A. R., *ibid.* **30**, 466 (1927).

[27] Hupfield, H., *Z. f. Phys.* **54**, 484 (1929).

[28] Kerschbaum, H., *Ann. d. Phys.* **79**, 465 (1926).

[29] —— *ibid.* **83**, 287 (1927).

[30] Koenig, H. D. and Ellett, A., *Phys. Rev.* **39**, 576 (1932).

[31] Kopfermann, H. and Tietze, W., *Z. f. Phys.* **56**, 604 (1929).

[32] Kuhn, W., *Danske Videnskabernes Selskab* (1926) (Zürich, Habilitationsschrift).

[33] Kunze, P., *Ann. d. Phys.* **85**, 1013 (1928).

[34] Ladenburg, R. and Senftleben, H., *Naturwiss.* **1**, 914 (1913).

[35] Ladenburg, R. and Reiche, F., *Ann. d. Phys.* **42**, 181 (1913).

[36] Ladenburg, R., *Verh. d. D. Phys. Ges.* **16**, 765 (1914).

[37] —— *Z. f. Phys.* **4**, 451 (1921).

[38] Ladenburg, R. and Minkowski, R., *ibid.* **6**, 153 (1921).

[39] Ladenburg, R., *ibid.* **48**, 15 (1928).

[40] Ladenburg, R. and Minkowski, R., *Ann. d. Phys.* **87**, 298 (1928).

[41] Ladenburg, R. and Wolfsohn, G., *Z. f. Phys.* **63**, 616 (1930).

[42] Ladenburg, R. and Levy, S., *ibid.* **65**, 189 (1930).

[43] Ladenburg, R. and Wolfsohn, G., *ibid.* **65**, 207 (1930).

[43a] Ladenburg, R. and Thiele, E., *ibid.* **72**, 697 (1931).

[44] Larché, K., *ibid.* **67**, 440 (1931).

[45] v. Malinowski, A., *Ann. d. Phys.* **44**, 935 (1914).

[46] Michels, W. C., *Phys. Rev.* **38**, 712 (1931).

[47] Milne, E. A., *Mon. Not. Roy. Ast. Soc.* **85**, 117 (1924).

[48] —— *Journ. Lond. Math. Soc.* **1**, 1 (1926).

[49] Minkowski, R., *Z. f. Phys.* **36**, 839 (1926).

[50] Minkowski, R. and Mühlenbruch, W., *ibid.* **63**, 198 (1930).

[51] Mrozowski, S., *Bull. Acad. Pol.* p. 464 (1930).

[52] Ornstein, S. and v. d. Held, E. F. M., *Ann. d. Phys.* **85**, 953 (1928).

[53] —— —— *Z. f. Phys.* **77**, 459 (1932).

[54] Orthmann, W., *Ann. d. Phys.* **78**, 601 (1925).

[55] Orthmann, W. and Pringsheim, P., *Z. f. Phys.* **43**, 9 (1927).

[56] Prokofjew, W. and Gamow, G., *ibid.* **44**, 887 (1927).

[57] Prokofjew, W. and Solowjew, W., *ibid.* **48**, 276 (1928).

[58] Prokofjew, W., *ibid.* **50**, 701 (1928).

[59] Puccianti, L., *Handb. d. Experimentalphysik*, **19**, 74.

[59a] Rabi, I. I. and Cohen, V., *Phys. Rev.* **43**, 582 (1933).

[60] Reiche, F., *Verh. d. D. Phys. Ges.* **15**, 3 (1913).

[61] Roschdestwensky, D., *Ann. d. Phys.* **39**, 307 (1912).

[62] —— *Trans. Opt. Inst. Len.* **2**, No. 13 (1921).

[63] Rump, W., *Z. f. Phys.* **29**, 196 (1924).

[64] Rupp, E., *Ann. d. Phys.* **80**, 528 (1926).

[64a] Schaffernicht, W., *Z. f. Phys.* **62**, 106 (1930).

[65] Schein, M., *Helv. Phys. Acta*, **2**, Suppl. 1 (1929).

[66] Schütz, W., *Z. f. Phys.* **45**, 30 (1927).

[67] —— *ibid.* **64**, 682 (1930).

[68] —— *Z. f. Astrophys.* **1**, 300 (1930).

[69] Slack, F. G., *Phys. Rev.* **28**, 1 (1926).

[70] Soleillet, P., *Compt. Rend.* **187**, 212 (1928).

[71] —— *ibid.* **194**, 783 (1932).

[72] Sugiura, Y., *Phil. Mag.* **4**, 495 (1927).

[73] Thomas, A. R., *Phys. Rev.* **35**, 1253 (1930).

[74] Tolman, R. C., *ibid.* **23**, 693 (1924).

[74a] Trumpy, B., *Z. f. Phys.* **66**, 720 (1930).

[74b] Van Atta, C. M. and Granath, L. P., *Phys. Rev.* **44**, 60 (1933).

[75] Voigt, W., *Münch. Ber.* p. 603 (1912).

[76] —— *Handb. d. Elekt. u. Mag. von Graetz*, **4**, 577 (1920).

[77] Vonwiller, O., *Phys. Rev.* **35**, 802 (1930).

[77a] Vinti, J. P., *ibid.* **42**, 632 (1932).

[78] Webb, H. W., *ibid.* **24**, 113 (1924).

[79] Webb, H. W. and Messenger, H. A., *ibid.* **33**, 319 (1929).

[80] Weiler, J., *Ann. d. Phys.* **1**, 361 (1929).

[81] Weingeroff, M., *Z. f. Phys.* **67**, 679 (1931).

[82] Weisskopf, V. and Wigner, E., *ibid.* **63**, 54 (1930).

[83] Weisskopf, V., *Phys. Zeits.* **34**, 1 (1933).

[83a] Wheeler, J. A., *Phys. Rev.* **43**, 258 (1933).

[84] Wien, W., *Ann. d. Phys.* **73**, 483 (1924).

[85] Wolfsohn, G., *Z. f. Phys.* **63**, 634 (1930).

[85a] —— *ibid.* **83**, 234 (1933).

[86] Zahn, H., *Verh. d. D. Phys. Ges.* **15**, 1203 (1913).

[87] Zehden, W. and Zemansky, M. W., *Z. f. Phys.* **72**, 442 (1931).

[88] Zehden, W., *Z. f. Phys.* **86** (1933).

[89] Zemansky, M. W., *Phys. Rev.* **36**, 219 (1930).

[90] —— *Z. f. Phys.* **72**, 587 (1931).

CHAPTER IV

COLLISION PROCESSES INVOLVING EXCITED ATOMS

1. TYPES OF COLLISION PROCESSES

On the basis of classical kinetic theory, a molecule was regarded as a rigid sphere with a definite radius, and a collision between two molecules was defined as an encounter in which the two spheres touched. It is of course no longer possible to ascribe a definite radius to a molecule or atom, and when two such bodies come together and part again, they do so in a manner which cannot be described in detail. It is therefore necessary to adopt a point of view which is independent of the actual shape and dimensions of the molecules, and which at the same time is unambiguous. This has been done with admirable clearness by Samson [84], and the following treatment will follow that of Samson rather closely.

1 *a*. THE MEANING OF "COLLISION". When any molecule passes another at any distance and with any relative velocity, a "collision" is said to take place. The best description of a particular collision is to give the relative velocity before the collision V and the perpendicular distance q between the centre of the second molecule and the line of the velocity V through the centre of the first. We may call this a (V, q) collision. For a (V, q) collision there is a probability $\phi(V, q)$ that a given process, say a transition of energy state, may occur on such a collision. We know nothing of the function $\phi(V, q)$ except that it must become zero for very large q and probably for very large V. In the case of an upward transition we know that $\phi(V, q)$ is zero until V reaches a value V_0 at which there is sufficient relative kinetic energy to produce the transition. There is no definite value of q for which $\phi(V, q)$ suddenly becomes zero and which could be called the sum of the radii of the two bodies. The values of $\phi(V, q)$ must also be expected to vary differently with V and q for different transition processes.

1*b*. THE MEANING OF "EFFECTIVE CROSS-SECTION". For
statistical purposes we resort to an artifice which gives a convenient index number with which to describe the statistical
average over all values of q. We can calculate easily the total
number of collisions of relative velocity V within a distance Q
say, denote it by $Z(V, Q^2) dV$ and, by equating this to the total
number of collisions actually known by experiment to produce
the given process, evaluate Q^2 and call it "the effective velocity
cross-section for the given process". (The cross-section area is
really πQ^2, but it is convenient to drop the π and refer simply to
Q^2 as the cross-section.) It will clearly be a function of velocity
and of the process considered and may have widely different
values for different processes. It has no relation whatsoever to
the gas-kinetic cross-section for the pair of molecules.

Since actual experiments usually do not differentiate between velocities but are carried on at a known temperature, it
is convenient to make a similar definition of an "effective
temperature cross-section for the given process". If $Z(T, \sigma^2)$
represents the total number of collisions of all velocities
within the distance σ at temperature T, then

$$Z(T, \sigma^2) = \int_0^\infty Z(V, Q^2) \, dV.$$

The left-hand member of the above equation can easily be
calculated on the basis of the Maxwellian distribution of
velocities along the lines of classical kinetic theory, and the
result is obtained

$$Z(T, \sigma^2) = 2Nn\sigma^2 \sqrt{2\pi RT \left(\frac{1}{M_1} + \frac{1}{M_2} \right)} \quad \ldots\ldots(91),$$

where M_1, M_2 and N, n are the molecular weights and concentrations respectively of the colliding molecules, and R is
the universal gas constant. "*The effective cross-section for the
process A*" is therefore calculated as follows: From the results
of an experiment in which it is known that collision processes
of type A occur, the number of such collisions per sec. per c.c.
is calculated. This number is equated to the right-hand
member of Eq. (91). The resulting value of σ^2 is denoted by
$\sigma_A{}^2$ and is called the effective cross-section for the process A.

1c. COLLISIONS OF THE SECOND KIND. Klein and Rosseland[37], in 1921, on the basis of thermodynamical reasoning, inferred that, if ionizing and exciting collisions take place in an assemblage of atoms and electrons, inverse processes must also take place, namely, collisions between excited atoms and electrons in which the excitation energy is transferred to the electrons in the form of kinetic energy. They called such collisions "collisions of the second kind". The term was extended by Franck and others to include collisions between excited atoms and normal atoms (or molecules) involving a transfer of the excitation energy from one to the other. As more and more types of collision processes were discovered the conception became broader, until now the expression "collision of the second kind" includes all collision processes in which the following conditions are fulfilled:

(1) One of the colliding particles is either an excited atom (metastable or otherwise) or an ion.

(2) The other colliding particle is either an electron, a normal atom or a normal molecule.

(3) During the collision *either all or a part* of the excitation of particle (1) is transferred to (2).

Typical examples of collisions of the second kind are given in Table XVIII. Some of these have been discussed in Chap. II, and the others will be considered throughout this chapter.

1d. PERTURBING COLLISIONS. There are types of collisions involving excited atoms in which either no energy or an extremely small amount of energy is transferred. Such collisions involve a perturbation of the excited atom so that its radiation or absorption characteristics are altered in some way. If the breadth of an absorption line of a gas is measured at high pressure or in the presence of a foreign gas, it is found to be greater than usual, indicating the existence of collisions which alter the absorbing characteristics of an atom. Such collisions can be called broadening collisions. If the percentage polarization of the resonance radiation emitted by a gas is measured in the presence of a foreign gas, it is found to be smaller than usual, indicating the existence of collisions which alter the position of the electric vector of the light emitted by an excited atom. Such collisions are called depolarizing collisions and will be treated in Chap. v.

TABLE XVIII

Particle No. (1)	Example of collision of second kind	Methods by which such collisions are studied
Atom in an excited state from which it can emit resonance radiation	$Cs\,(^2P) + e$ $= Cs\,(6\,^2S_{1/2}) + e$	Optical and electrical measurements on the positive column of a gas discharge
	$Hg\,(6\,^3P_1) + H_2$ $= Hg\,(6\,^1S_0) + H + H$	Quenching of resonance radiation. Reduction of H_2 pressure as oxide is reduced by H atoms
	$Hg\,(6\,^3P_1) + Na$ $= Hg\,(6\,^1S_0) + Na\,(10\,^2S_{1/2})$	Sensitized fluorescence
	$Hg\,(6\,^3P_1) + N_2$ $= Hg\,(6\,^3P_0) + (N_2)'$	Quenching of resonance radiation. Rapidity of escape of resonance radiation
	$Na\,(3\,^2P_{3/2}) + A$ $= Na\,(3\,^2P_{1/2}) + A$	Radiation of D_1 line when excited by D_2 in the presence of Argon
	$Cs\,(m\,^2P) + Cs$ $= (CsCs)^+ + e$	Photo-ionization of vapour by means of $6\,^2S_{1/2} - m\,^2P$ at various pressures
	$Tl\,(7\,^2S_{1/2}) + I_2$ $= (TlI)' + I'$	Quenching of Tl line 5351 when excited Tl atoms are produced by photo-dissociation of TlI
Metastable atom	$Hg\,(6\,^3P_0) + H_2$ $= (HgH)' + H$	Bands in fluorescence
	$Hg\,(6\,^3P_0) + N_2$ $= Hg\,(6\,^3P_1) + N_2$	Rapidity of escape of resonance radiation
Ion	$Ne^+ + Cu$ $= (Cu^+)' + Ne$	Enhancement of spark lines in gas discharge
	$He^+ + N_2$ $= He + N_2^+$	Absorption of canal rays

Note. A dash means an unspecified excited state. A chemical symbol unaccompanied by any other designation refers to the normal state.

2. CLASSICAL THEORY OF LORENTZ BROADENING OF AN ABSORPTION LINE

2*a*. THE PHENOMENON OF LORENTZ BROADENING. The interpretation of the broadening of spectral lines as due to collisions goes back to Rayleigh and Helmholtz. The first experimental evidence of the pressure broadening of spectral lines was obtained by Michelson [58], who showed that emission lines were broadened by an increase of pressure and who worked out an expression for the frequency distribution of the emitted radiation. Schönrock [85] extended Michelson's theory of emission lines and obtained an expression for the half-breadth of an emission line in terms of the mean free path and the temperature. In view of the complicated conditions that are present in a source of light such as an arc or spark discharge, no simple expression for the frequency distribution or half-breadth of an emission line can hope to take into account all the broadening factors. It is therefore much more fruitful to consider the broadening of an absorption line, since the conditions inside of an absorption tube can be made relatively simple.

Lorentz [51] was the first to formulate a simple theory of the pressure broadening of absorption lines, and was able to calculate both the half-breadth and the frequency distribution of an absorption line that was broadened by collisions *either* with other absorbing atoms *or* with foreign gas molecules. Since the time of Lorentz many experiments have been performed on the broadening of an absorption line by foreign gases, and most of these experiments, notably those of Füchtbauer and his co-workers, show the following characteristics as the foreign gas pressure is increased:

(1) The absorption line is broadened.

(2) The maximum of the absorption line is shifted.

(3) The absorption line becomes asymmetrical.

These three phenomena are illustrated in Fig. 30.

The simple theory of Lorentz is capable of giving an interpretation, on the basis of classical theory, of only the first of these effects. Both the shift and the asymmetry are apparently outside the realm of classical theory and require for their explanation the introduction of quantum ideas. The quantum theory of Lorentz broadening along with quantum explanations of shift and asymmetry will be touched upon later. For an approximate interpretation of existing experimental data the simple Lorentz theory which will be given in the next paragraph will be found to be helpful.

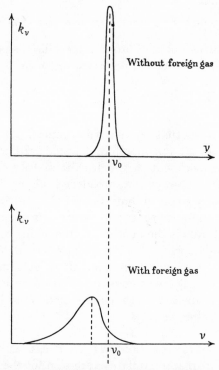

Fig. 30. Lorentz broadening of an absorption line, showing broadening, shift and asymmetry. (Exaggerated.)

2b. THE SIMPLE LORENTZ THEORY. The effect of collisions upon the absorbing and emitting characteristics of a classical "oscillator-atom" was treated by Lorentz in a manner analogous to that of radiation damping. If an absorbing atom performs Z collisions per second, with the molecules of a foreign gas, the resulting effect is equivalent to Z interruptions per second in an undamped wave train. The expression for the absorption coefficient of a gas under these conditions was found by Lorentz to be of the same form as that when natural broadening is present, namely

$$k_\nu = \frac{\text{const.}}{1 + \left[\dfrac{2(\nu - \nu_0)}{\Delta\nu_L}\right]^2} \qquad \ldots\ldots(92),$$

where $\Delta\nu_L$ is called the *Lorentz half-breadth* and is given by

$$\Delta\nu_L = \frac{1}{\pi} \left(\begin{array}{c}\text{number of broadening collisions per} \\ \text{second per absorbing atom}\end{array}\right)$$

$$= \frac{Z_L}{\pi} \qquad \ldots\ldots(93).$$

It is quite apparent that Eq. (92) is a symmetrical curve with a maximum at $\nu = \nu_0$, and therefore it would seem incapable of handling an experiment in which shift and asymmetry are present. When, however, a broadened absorption line is spectroscopically resolved, both the shift and the asymmetry can be corrected for, so that the Lorentz half-breadth may be obtained, and from it the number of broadening collisions. In other experiments where the line is not resolved, the effect of the shift can be eliminated entirely, leaving asymmetry as the only error. Eq. (92) therefore is by no means useless, but can be used in conjunction with some experiments to yield an approximate value for $\Delta\nu_L$.

2c. COMBINATION OF LORENTZ, NATURAL AND DOPPLER BROADENING. An approximate expression for the absorption coefficient of a gas under conditions in which Lorentz, natural, and Doppler broadening are present can be obtained by a method due to Reiche[81], in which the interdependence of Doppler and Lorentz broadening is ignored, i.e. as if the collisions performed by one absorbing atom with the foreign gas molecules did not take place irregularly, but with a certain constant time interval equal to $1/Z_L$. With this simplification Lorentz broadening can be combined with Doppler broadening in exactly the same way that natural broadening was com-

bined with Doppler broadening. The resulting absorption
coefficient is given by

$$k_\nu = \frac{2k_0}{\pi\,(\Delta\nu_N + \Delta\nu_L)} \int_{-\infty}^{\infty} \frac{\exp-\left[\dfrac{2\delta}{\Delta\nu_D}\sqrt{\ln 2}\right]^2}{1+\left[\dfrac{2}{\Delta\nu_N+\Delta\nu_L}(\nu-\nu_0-\delta)\right]^2}\,d\delta \dots (94),$$

where k_0 is the maximum absorption coefficient when *only*
Doppler broadening is present, $\Delta\nu_N$ is the natural half-breadth
of the line and $\Delta\nu_D$ is the Doppler breadth. It has been mentioned before that Voigt[99] calculated on the basis of classical
theory the absorption coefficient of a gas under conditions in
which Doppler broadening, natural damping and any other
damping process were present. This theory is capable of yielding a more accurate expression than Eq. (94), because Lorentz
broadening can be introduced as a damping term which is not
constant, but is instead a function of the velocity of the
absorbing atom. When this is done, an integral is obtained
which is much more complicated than Eq. (94). It was pointed
out by Reiche, in a private communication, that Voigt's integral represents a symmetrical function of ν with a maximum
at ν_0, and therefore, since it does not account for the shift and
asymmetry, its slight advantage over Eq. (94) hardly warrants·
its use, in view of the added mathematical difficulties.

In dealing with the pressure broadening of infra-red absorption lines, Dennison[13] derived a formula which is substantially
the same as Voigt's general formula, but which, for purposes of
calculation, had to be replaced by two simple Lorentz expressions: one for the centre of the line and one for the edges. It is
doubtful whether Dennison's final formulas are very much
more accurate than Eq. (94), in that they also do not take into
account the asymmetry and shift that are present. For a
simple derivation of Eq. (94) on the basis of classical theory the
reader is referred to a recent paper by Weisskopf[104]. Introducing the quantities

$$\omega = \frac{2\,(\nu-\nu_0)}{\Delta\nu_D}\sqrt{\ln 2} \qquad \dots\dots(95)$$

and $\qquad a' = \dfrac{\Delta\nu_N+\Delta\nu_L}{\Delta\nu_D}\sqrt{\ln 2} \qquad \dots\dots(96),$

and letting $y = \dfrac{2\delta}{\Delta \nu_D} \sqrt{\ln 2}$,

Eq. (94) becomes $k_\nu = k_0 \dfrac{a'}{\pi} \displaystyle\int_{-\infty}^{\infty} \dfrac{e^{-y^2}\, dy}{a'^2 + (\omega - y)^2}$ ......(97).

The above equation will be recognized to be of the same form as that representing the absorption coefficient when natural and Doppler broadening only are present (see Chap. III, §2a). There is, however, an important distinction. The quantity a' cannot, in this case, be treated as a small quantity; for, due to the presence of $\Delta \nu_L$, it may be made as large as we please by increasing the foreign gas pressure. The integral, therefore, in Eq. (97) cannot be expressed in a simple form, but must be evaluated by series or by numerical integration. A table of values of k_ν/k_0 for several values of a' and ω will be found in the Appendix. In Fig. 31, Eq. (97) is plotted as a function of ω for four different values of a'.

3. EXPERIMENTS ON LORENTZ BROADENING

3a. PHOTOGRAPHIC MEASUREMENTS. A most extensive investigation of Lorentz broadening was made by Füchtbauer, Joos and Dinkelacker [22] on the mercury resonance line 2537. A beam of light from a source emitting a continuous spectrum was sent through an absorption tube containing a mixture of mercury vapour at room temperature and a foreign gas. The light was then focused on the slit of a spectrograph, and photographs were taken of the 2537 line in absorption, while the mercury vapour pressure remained constant and the foreign gas pressure was increased from 10 to 50 atmospheres. So much broadening was produced at these high foreign gas pressures that the absorption line was easily resolved by the slit of the spectrograph. Under these conditions also, the quantity a' in Eq. (97) is so large that the absorption coefficient reduces to the simple Lorentz form

$$k_\nu = \frac{\text{const.}}{1 + \left[\dfrac{2(\nu - \nu_0)}{\Delta \nu_L} \right]^2}.$$

The experimental curves of absorption coefficient against

frequency, obtained by taking photometer measurements of the plates, showed the shift and asymmetry already mentioned. The half-breadth of the shifted curve was taken to be the Lorentz half-breadth. In this way $\Delta\nu_L$ was measured with

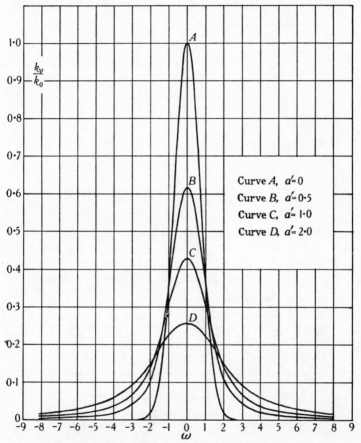

Curve A, $a'=0$
Curve B, $a'=0.5$
Curve C, $a'=1.0$
Curve D, $a'=2.0$

Fig. 31. Simple Lorentz broadening of a Doppler line.

various foreign gases at various pressures, and it was found that the graph of $\Delta\nu_L$ against p (the foreign gas pressure) was a straight line with a different slope for each foreign gas. A few of these curves are shown in Fig. 32. The results will be interpreted in the light of effective cross-sections later on in this chapter.

3*b*. MEASUREMENTS INVOLVING MAGNETO-ROTATION. The conditions under which Schütz[87] studied the Lorentz broadening of the sodium resonance lines were the same as those described in Chap. III, § 5*b*. An absorption tube containing sodium vapour and a foreign gas was placed in a longitudinal

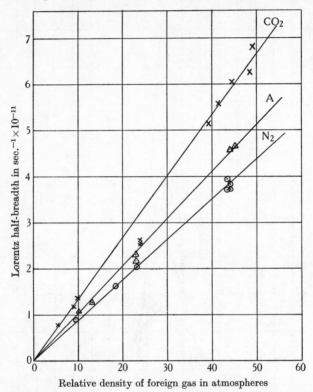

Fig. 32. Füchtbauer, Joos and Dinkelacker's experiments on Lorentz broadening of the mercury resonance line.

magnetic field and between a polarizing and an analysing Nicol. A beam of light from a source emitting a continuous spectrum was sent through the tube and focused on the slit of a spectroscope-photometer arrangement. With zero magnetic field and crossed Nicols, no light entered the photometer; but with the Nicols kept crossed and the magnetic field established, magneto-rotation at the edges of the absorption line caused light to enter the photometer. It has already been shown that the amount of light of a particular frequency passing through the analysing Nicol depends upon the absorption coefficient for that frequency, the strength of the magnetic field, and certain atomic constants dealing with the Zeeman effect. The sodium vapour pressure was high enough to absorb completely the centre of the line, and the foreign gas pressure was kept very low so that a' was kept small. Under these conditions, Eq. (97) reduces to

$$k_\nu = \text{const.} \frac{\Delta \nu_N + \Delta \nu_L}{(\nu - \nu_0)^2} \qquad \ldots \ldots (98),$$

which, when expressed in classical notation, and introduced into the equations of Chap. III, § 5b, leaves them completely unaltered in form. Only the quantity ν' is changed, being now the sum of natural and Lorentz damping. Since the final equation is a relation between the intensity of light entering the photometer and ν', the method of Schütz allows ν' to be obtained at any foreign gas pressure. The results of Schütz's experiments were expressed in the form of a graph between ν'/ν_0' and the foreign gas pressure, where ν_0' represents the natural damping. Some of these curves are shown in Fig. 33, and they are seen to be straight lines. Their interpretation will be given later.

3c. EXPERIMENTS ON THE ABSORPTION OF RESONANCE RADIATION. It will be remembered that the absorption of a beam of resonance radiation in traversing a gas depends upon the form (frequency distribution) of the emission and absorption lines. Orthmann[76] was the first to study the Lorentz broadening of the mercury resonance line by measuring the absorption of a beam of mercury resonance radiation by a mixture of mercury vapour and hydrogen. Neumann[71] extended Orthmann's measurements, using argon, air, helium and hydrogen. In later years Kunze[42] and Zemansky[114] performed the same experiment under conditions which allowed the Lorentz half-breadth to be calculated with fair accuracy. The theory of the method is very simple. Assuming the mercury resonance line to consist of five equal, completely separate, hyperfine structure components; and representing the emission line from the resonance lamp by the expression E_ν, and the absorption coefficient of the mixture of mercury vapour and

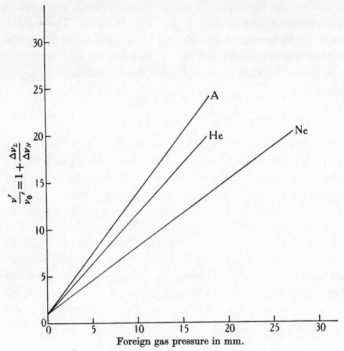

Fig. 33. Schütz's experiments on Lorentz broadening of sodium resonance line.

foreign gas, by k_ν, the absorption A is given by (see Chap. III, §4b)

$$A = \frac{\int_0^\infty E_\nu (1 - e^{-k_\nu l})\, d\nu}{\int_0^\infty E_\nu\, d\nu} \qquad \ldots\ldots(99).$$

In the experiments of Kunze, the foreign gas was admitted to the resonance lamp at exactly the same pressure as in the absorption cell, thereby eliminating any error due to the shift of the absorption line relative to the emission line. This necessitated the use of inert gases only, since a gas like hydrogen or oxygen, if admitted to a resonance lamp, would quench the resonance radiation to such a low intensity that it could not be measured. Since the resonance lamp employed by Kunze satisfied fairly well the conditions laid down in Chap. III, § 3 a for an ideal resonance lamp, and since the thickness of the emitting layer of the resonance lamp was nearly equal to the thickness of the absorption cell, the expression for E_ν becomes [see Eq. (50)]

$$E_\nu = \text{const.} \, (1 - e^{-k_\nu l}) \qquad \ldots \ldots (100),$$

and A can be written

$$A = \frac{\displaystyle\int_0^\infty (1 - e^{-\frac{k_\nu}{k_0} k_0 l})^2 \, d\nu}{\displaystyle\int_0^\infty (1 - e^{-\frac{k_\nu}{k_0} k_0 l}) \, d\nu} \qquad \ldots \ldots (101).$$

The conditions of temperature and mercury vapour pressure in Kunze's experiment were such that $k_0 l$ was 0·475. Substituting this value for $k_0 l$ in the above formula, and replacing k_ν / k_0 by the expression given by Eq. (97), a result is obtained which can be integrated graphically with the aid of the table in the Appendix. (The graphical integration is to be done for all five hyperfine structure components.) The result is a different value of A for all the different values of a'. Since a' contains the ratio $\Delta\nu_L / \Delta\nu_D$, the theoretical curve of A against a' can be used to give the value of $\Delta\nu_L / \Delta\nu_D$ corresponding to any experimentally measured value of A at any foreign gas pressure. It is to be expected, of course, that the neglect of the accurate hyperfine structure of the line and its asymmetry in broadening will produce a small error in the final value of $\Delta\nu_L$. In this way, the ratio $\Delta\nu_L / \Delta\nu_D$ was obtained for the three inert gases as a function of the gas pressure. A graph of the results is shown in Fig. 34.

In Zemansky's experiments, many gases were used which could not be introduced into the resonance lamp because of their strong quenching ability, and consequently the emission line remained fixed while the absorption line was shifted and broadened by the foreign gases. It will be seen later, however, that the results agree quite well with those of Füchtbauer, Joos

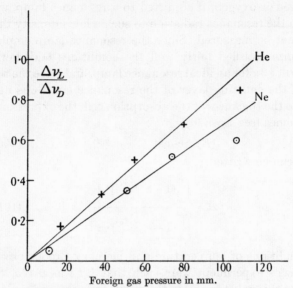

Fig. 34. Kunze's experiments on Lorentz broadening of mercury resonance line.

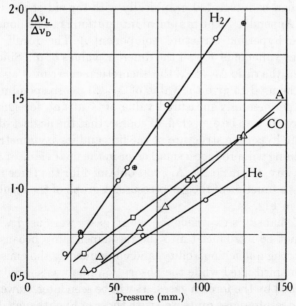

Fig. 35. Zemansky's experiments on Lorentz broadening of mercury resonance line.

and Dinkelacker and with those of Kunze. To obtain a theoretical curve between A and a' to be used with Zemansky's experimental results, the emission line was represented by the expression [see Eq. (60)]

$$E_\nu = \text{const.} \, e^{-\left(\frac{\omega}{1\cdot21}\right)^2} \qquad \ldots\ldots(102),$$

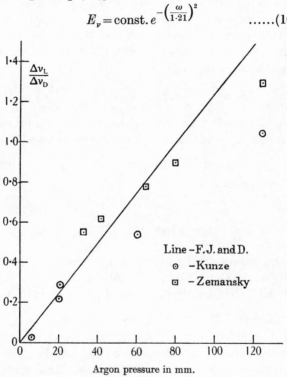

Fig. 36. Lorentz broadening of mercury resonance line by argon.

and the absorption became

$$A = \frac{\int_{-\infty}^{\infty} e^{-\left(\frac{\omega}{1\cdot21}\right)^2}(1 - e^{-\frac{k_\nu}{k_0}k_0 l})\, d\omega}{\int_{-\infty}^{\infty} e^{-\left(\frac{\omega}{1\cdot21}\right)^2}\, d\omega} \qquad \ldots\ldots(103).$$

In his experiments, $k_0 l$ was $4\cdot44$ and, upon introducing Eq. (97) for k_ν/k_0 and integrating graphically, A was calculated as a function of a'. The results for a few foreign gases are shown in Fig. 35; and in Fig. 36, a comparison of the results of Füchtbauer, Joos and Dinkelacker, Kunze, and Zemansky is shown for the case of argon.

3d. EVALUATION OF EFFECTIVE CROSS-SECTIONS FOR LORENTZ BROADENING. It is seen from Figs. 33, 34, 35 and 36, that all experiments on Lorentz broadening provide linear relations between the Lorentz half-breadth and the foreign gas pressure. This is in agreement with the Lorentz formula for $\Delta \nu_L$ given by Eqs. (93) and (91), namely

$$\Delta \nu_L = \frac{Z_L}{\pi} = \frac{2}{\pi} \sigma_L^2 N \sqrt{2\pi RT \left(\frac{1}{M_1} + \frac{1}{M_2} \right)} \quad \dots (104),$$

where σ_L^2 is the effective cross-section for Lorentz broadening and N is the number of foreign gas molecules per c.c. If p is the foreign gas pressure in mm., then $N = \dfrac{9740}{T} \cdot p \times 10^{15}$, and Eq. (104) becomes

$$\frac{\dfrac{\Delta \nu_L}{\Delta \nu_D}}{p} = \frac{1}{\Delta \nu_D} \left\{ 1 \cdot 95 \times 10^{19} \sqrt{\frac{2R}{\pi T} \left(\frac{1}{M_1} + \frac{1}{M_2} \right)} \right\} \cdot \sigma_L^2 \dots (105),$$

which shows how the effective cross-section σ_L^2 may be obtained from the slope of any curve in Figs. 33 to 36.

Table XIX contains all the values of σ_L^2 obtained in this way.

3e. LORENTZ BROADENING IN A SODIUM FLAME. The increase in intensity of a sodium flame, as the thickness of the flame or the sodium concentration, or both, were increased, was first measured by Gouy [25] in 1879. Since then, similar experiments have been made by Senftleben [90], Wilson [106], Locher [48], Child [11] and Bonner [10], who, with the exception of Child, showed that the intensity of the light emitted by the flame was a function of the product of flame thickness and the concentration of the sodium salt that is sprayed into the flame.

When the partial pressure of the sodium vapour in the flame is small, the flame can be treated in exactly the same manner as the emitting layer of an ideal resonance lamp (see Chap. III, § 3a). If I_1 denotes the intensity of light (comprised entirely of the D lines) emitted by a flame of thickness l, then we have

$$I_1 = \text{const.} \int (1 - e^{-k_\nu l}) \, d\nu \quad \dots \dots (106),$$

where k_ν depends on the amount of sodium vapour present, and the absorption line *form*, which is obviously determined by natural and Lorentz damping and the Doppler effect. (Hyperfine structure is practically wiped out by Lorentz broadening.) A convenient way of representing the experimental results is to plot the ratio of the intensity from a flame of thickness

TABLE XIX

Absorbing gas	Foreign gas	$\sigma_L^2 \times 10^{16}$ Füchtbauer, Joos and Dinkelacker	$\sigma_L^2 \times 10^{16}$ Zemansky	$\sigma_L^2 \times 10^{16}$ Kunze	$\sigma_L^2 \times 10^{16}$ Schütz
Hg	He	—	15·0	21·4	—
,,	H_2	27·8	24·5	—	—
,,	Ne	—	—·	35·7	—
,,	CO	—	44·5	—	—
,,	N_2	64·8	51·0	—	—
,,	O_2	65·1	—	—	—
,,	CH_4	—	42·8	—	—
,,	H_2O	68·5	—	—	—
,,	A	88·9	61·5	62·0	—
,,	CO_2	125	—	—	—
,,	NH_3	—	71·2	—	—
,,	C_3H_8	—	73·5	—	—
Na	He	—	—	—·	31·4
,,	Ne	—	—	—	37·8
,,	H_2	—	—	—	33·6
,,	N_2	—	—	—	68·9
,,	A	—	—	—	81·0

$2l\,(I_2)$ to that from a flame of thickness $l\,(I_1)$ against I_1. This ratio can be written

$$\frac{I_2}{I_1} = \frac{\int(1 - e^{-2k_0 l \frac{k_\nu}{k_0}})\,d\nu}{\int(1 - e^{-k_0 l \frac{k_\nu}{k_0}})\,d\nu} \qquad \ldots\ldots(107),$$

and can be evaluated for various values of $k_0 l$ (and therefore of I_1) once k_ν/k_0 is known as a function of ν. In Fig. 37 are shown the experimental results of Gouy and Bonner, with the ratio I_2/I_1 plotted against I_1. The abscissa scales are chosen to make the two sets of results coincide with each other as much as possible. It is quite clear that I_2/I_1 first attains a minimum value of about 1·35 and then rises slowly.

Schütz[88] was the first to obtain a curve of this shape theoretically. He introduced into Eq. (107) an expression for k_ν/k_0

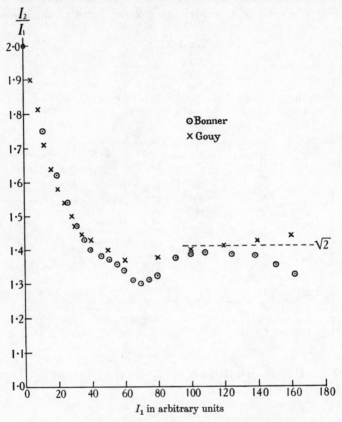

Fig. 37. Experiments of Gouy and Bonner on the emission of sodium flames.

equivalent to our Eq. (97), representing the absorption coefficient when natural and Lorentz damping and Doppler effect are present. With a' equal to any arbitrarily chosen value, Schütz was able to show that I_2/I_1 first attained a minimum and then rose slowly to the value $\sqrt{2}$. He estimated that, to agree with Gouy and Senftleben's results, a' would have to be about 0·5. From this we can estimate the effective cross-section associated with sodium and air.

Since
$$a' = \frac{\Delta\nu_N + \Delta\nu_L}{\Delta\nu_D} \sqrt{\ln 2} = 0\cdot5,$$

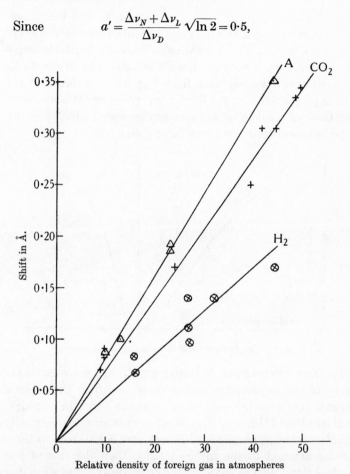

Fig. 38. Füchtbauer, Joos and Dinkelacker's measurements of the shift of the mercury resonance line.

then, since $\Delta\nu_N/\Delta\nu_D$ is very small, $\Delta\nu_L/\Delta\nu_D = 0\cdot6$. Replacing $\Delta\nu_L$ by Z_L/π and Z_L by its gas kinetic expression, and calculating $\Delta\nu_D$, $\sigma_L{}^2$ is found to be roughly 15×10^{-16} cm.^{2}, which is considerably smaller than Schütz's value for nitrogen alone. A more recent measurement of a' was made by van der Held and Ornstein[28] for a sodium flame. The resulting value of a', 0·53, is in good agreement with that calculated by Schütz.

3*f*. THE SHIFT OF THE ABSORPTION LINE. The most complete experiments on the shift of a Lorentz-broadened absorption line were made by Füchtbauer, Joos and Dinkelacker[22] on the mercury resonance line. The shift was found to be always toward the red, and, from Fig. 38, it is seen that the shift is proportional to the foreign gas pressure. It is apparent also that, in a rough way, those foreign gases which produce large broadening also produce large shifts.

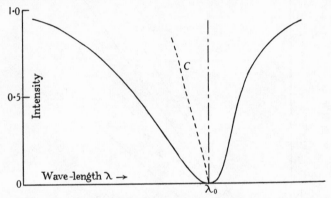

Fig. 39. Graphical method of indicating asymmetry.

3*g*. THE ASYMMETRY IN BROADENING. The most extensive study of the asymmetry in the frequency distribution of a Lorentz-broadened absorption line was made by Minkowski[61] on the sodium D lines. If Fig. 39 represents an asymmetrically broadened absorption line, with ordinates equal to the ratio of the transmitted to the incident light, then the curved line marked C is drawn by dividing the ordinate axis into a number of parts and indicating at each level the centre of the line. In this way the asymmetry of a sodium resonance line broadened by a number of gases is indicated in Fig. 40. It is seen that, with H_2 and He, the asymmetry is toward the violet, and with all the other gases the asymmetry is toward the red. It is also apparent that, in a rough way, those gases which produce the most asymmetry are also those which broaden the most. These results were explained by Minkowski in a qualitative way by assuming that there is an interchange of kinetic energy and excitation energy during collision, whereby an atom, capable of absorbing the frequency ν_0, can absorb a frequency *smaller* than ν_0 plus a small amount of the kinetic energy of the collision. Or vice versa: an atom, capable of absorbing the frequency ν_0, can absorb a frequency *larger* than ν_0 and at the same time give up the remainder to the kinetic energy of the collision. These ideas are in agreement with similar ideas advanced by Oldenberg [74] to explain the appearance of bands that were emitted by a mixture of mercury vapour and some

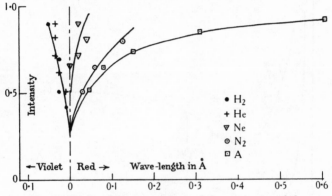

Fig. 40. Minkowski's measurements of the asymmetry of the sodium resonance line.

of the inert gases. They are also contained implicitly in the general theory of Lorentz broadening developed by Weisskopf and extended by Lenz, which will be discussed in the next section.

4. QUANTUM THEORY OF LORENTZ BROADENING

4*a*. PRELIMINARY THEORIES. The first attempt at a quantum theory of Lorentz broadening was made by Jablonski [31]. He considered a collision between an absorbing atom and a foreign gas molecule as a temporary formation of a quasi-molecule. In Fig. 41 are shown the Franck-Condon curves of the two states of this molecule, the first state corresponding to the molecule : normal absorbing atom A plus foreign gas molecule F, and the second state corresponding to the molecule : excited absorbing atom A' plus foreign gas molecule F. The difference between the two Franck-Condon curves determines the frequency absorbed (or emitted) as a function of the separation r. Jablonski's ideas were only qualitative and were not capable of yielding an expression for the frequency distribution of the absorption line, or for the half-breadth.

The next important step in the development of an accurate theory of Lorentz broadening was made by Margenau [53, 54] in

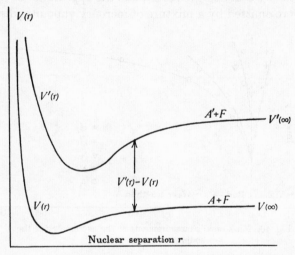

Fig. 41. Franck-Condon curves referring to collision broadening.

America and at about the same time by Kulp [41] in Germany. These authors extended the ideas of Jablonski, and, applying the statistical theory of density fluctuations, were able to calculate the shift of an absorption line in terms of the pressure of the foreign gas and of atomic constants. The shift was shown to be proportional to the pressure, in agreement with experiment. London's theory of van der Waals' forces was used by both authors, which is tantamount to considering only the right-hand portions of the Franck-Condon curves (at large values of r) where the mutual potential energy of the quasi-molecule is due to polarization, or $V'(r) - V(r) = -Cr^{-6}$.

4b. WEISSKOPF'S THEORY. The first really successful theory of Lorentz broadening was developed by Weisskopf[105], who showed the quantum-mechanical principles involved, and gave the mathematical tools (which, by the way, are identical with those used in classical electron theory) for calculating the frequency distribution of the broadened line. According to Weisskopf, Lorentz broadening is regarded as a conversion of translational energy into light energy and vice versa, according to the ideas of Minkowski[61] and Oldenberg[74]. Wave mechanically, it is analogous to the electron-vibration bands of a diatomic molecule in which the energy of motion of the nuclei can be either added to or subtracted from the electron terms. In the case of the diatomic molecule, the spectrum is discrete, because of the *regular* character of the nuclear motions. In the case of Lorentz broadening, however, the two sides of the broadened line represent continuous spectra due to the irregularity of the motions involved. The process of Lorentz broadening is to be considered as the quantum analogue of a change in the frequency of vibration of an oscillator atom caused by the approach of a perturbing foreign gas atom, so that the phases of the normal vibration before and after the collision no longer agree. This idea was first put forward by Lenz[46]. If we imagine the absorbing atom stationary, and the perturbing atom flying past along a line which is at a distance r from the absorbing atom, then the phase change will depend upon r. The distance of closest approach, ρ, was defined by Weisskopf as that value of r for which the phase change is 1. Replacing the whole Franck-Condon curves by their right-hand portions, and using London's expression for the potential energy due to polarization,

$$V'(r) - V(r) = -Cr^{-6} \qquad \ldots\ldots(108),$$

ρ was found to be approximately

$$\rho \doteq \sqrt[5]{\frac{C}{c_0}} \qquad \ldots\ldots(109),$$

where c_0 is the relative velocity of the colliding partners.

London has given a rigorous expression for C, but it is more convenient to calculate the maximum and minimum possible values of C for purposes of comparison with experiment. From London's formula

$$C_{\text{max.}} = \tfrac{3}{2}\alpha_2 \Delta F (\alpha_1 + \alpha') \qquad \ldots\ldots(110),$$

$$C_{\text{min.}} = \tfrac{3}{2}\alpha_2 \Delta F \left(\alpha_1 \frac{\Delta E'}{\Delta F + \Delta E''} + \alpha'\right) \quad \ldots\ldots(111),$$

and $\qquad \alpha' = \dfrac{e}{m} \left(\dfrac{h}{2\pi}\right)^2 \dfrac{f_0 \Delta F}{(\Delta F - \Delta E_0)(\Delta E_0)^2} \ldots\ldots(112),$

where α_1 is the polarizability of *excited* atom A,

$\qquad \alpha_2$ is the polarizability of normal foreign molecule F,

$\qquad f_0$ is the f-value of the line in question,

$\Delta E_0 = h\nu_0$,

$\Delta F = $ mean excitation energy of F,

$\Delta E' = $ energy difference between excited state of atom A and the next highest combining state,

$\Delta E'' = $ energy difference between excited state of atom A and ionization.

The upper and lower limits of C can be calculated from data in Table XX.

TABLE XX

Atom or molecule	Polarizability $\alpha \times 10^{24}$	Smallest excitation energy in volts	Ionization energy in volts
He	0·20	19·8	24·6
A	1·63	11·5	15·6
Ne	0·39	16·5	21·5
N_2	1·74	6·5	17
O_2	1·57	8	13
CO_2	2·7	—	14·3
CO	1·9	6·4	10
Na (3^2P)	96	2·1	5·1
Hg (6^3P_1)	20	4·7	10·4

A comparison of values of ρ calculated by means of Weisskopf's formulas, and experimental values of σ_L, are given in Tables XXI and XXII, which have been taken from Weisskopf's report in the *Physikalische Zeitschrift* [105].

Two facts are immediately evident from the tables. First, there is a fair agreement between the theoretical values of ρ and the experimental measurements of σ_L. Second, there is *no* clear-cut relation between the ability of a foreign gas molecule to broaden a line and to take energy from the excited atom, because σ_L and σ_Q (σ_Q refers to quenching of resonance radiation) are quite different. To the agreement between ρ and σ_L very little importance can be attached. σ_L was obtained from either direct or indirect measurements of the Lorentz half-

TABLE XXI

BROADENING OF MERCURY RESONANCE LINE

Foreign gas molecule	Füchtbauer, Joos and Dinkelacker $\sigma_L \times 10^8$	Zemansky $\sigma_L \times 10^8$	Kunze $\sigma_L \times 10^8$	Theoretical max. $\rho \times 10^8$	Theoretical min. $\rho \times 10^8$	Quenching $\sigma_Q \times 10^8$
H_2	5·27	4·95	—	—	—	2·94
He	—	3·88	4·63	5·0	3·2	0
A	9·44	7·85	7·88	8·3	5·4	0
Ne	—	—	5·98	6·7	4·4	0
N_2	8·05	7·15	—	7·9	5·1	0·524
O_2	8·07	—	—	—	—	4·46
CO_2	11·2	—	—	9·1	6·0	1·88
CO	—	6·68	—	—	—	2·42

TABLE XXII

BROADENING OF SODIUM RESONANCE LINE

Foreign gas molecule	Schütz $\sigma_L \times 10^8$	Minkowski $\sigma_L \times 10^8$	Theoretical max. $\rho \times 10^8$	Theoretical min. $\rho \times 10^8$	Quenching $\sigma_Q \times 10^8$
H_2	5·8	—	—	—	2·52
He	5·6	—	6·3	4·8	0
A	9·0	7·9	10·0	7·6	0
Ne	6·15	—	7·9	6·0	0
N_2	8·3	7·7	—	—	3·09

breadth through the agency of a specific relation connecting σ_L and the half-breadth. In Weisskopf's theory, however, no expression was obtained for the half-breadth, and therefore it is not clear what quantity should be compared with ρ.

4c. LENZ'S THEORY. This point was finally cleared up by Lenz [47] in a very important paper which carries the theory of Lorentz broadening to a point much farther than it was brought by Weisskopf. On the basis of a more general expression for the mutual potential energy of the quasi-molecule formed by the colliding partners, namely

$$V'(r) - V(r) = -ar^{-p} \qquad \ldots\ldots(113),$$

Lenz was able to obtain an analytic expression for the frequency distribution of a line which contained all three known phenomena—broadening, asymmetry and shift; and also expressions for the half-breadth and shift in terms of atomic constants and a distance ρ_0 defined as the distance of closest approach. The distance of closest approach, ρ_0, is defined as the distance from an absorbing atom to the line representing the relative velocity of the absorbing and perturbing atoms, at which a phase change of 2π occurs. Calling the relative velocity c_0, this distance is found to be

$$\rho_0 = 1\cdot 50 \left(\frac{a}{c_0}\right)^{\frac{1}{p-1}} \qquad \ldots\ldots(114),$$

from which ρ_0 can be calculated once the constants a and p are known. It should be remarked that Lenz's definition of ρ_0 differs from Weisskopf's definition of ρ to the extent that

$$\rho_0 = 1\cdot 5\,(2\pi)^{\frac{1}{p-1}}.\rho \qquad \ldots\ldots(115).$$

Instead of assuming $p = 6$ and attempting to determine the value of a in order to compute ρ_0, Lenz determined p from experimental measurements of shift and half-breadth, and ρ_0 from measurements of half-breadth, in the following way. The shift of the line, in sec.$^{-1}$, was shown to be

$$\Delta\nu_S = -n c_0 \rho_0^{\,2} \frac{2\sqrt{\pi}\,(2\kappa\mu)^{\frac{2}{p-1}}}{2\cdot 25\,(p-1)} \cdot \frac{\Gamma\left(2 - \dfrac{1}{p-1}\right)}{\Gamma\left(1 + \dfrac{2}{p-1}\right)} \cdot \frac{\sin\dfrac{\pi}{p-1}}{\sin\dfrac{2\pi}{p-1}} \quad \ldots(116),$$

and the half-breadth, also in sec.$^{-1}$, to be

$$\Delta\nu_L = 2n c_0 \rho_0^{\,2} \frac{2\sqrt{\pi}\,(2\kappa\mu)^{\frac{2}{p-1}}}{2\cdot 25\,(p-1)} \cdot \frac{\Gamma\left(2 - \dfrac{1}{p-1}\right)}{\Gamma\left(1 + \dfrac{2}{p-1}\right)} \cdot \frac{\cos\dfrac{\pi}{p-1}}{\sin\dfrac{2\pi}{p-1}} \quad \ldots(117),$$

where n is the number of perturbing (foreign gas) atoms per c.c.,

c_0 is the relative velocity of the absorbing and perturbing atoms, and the quantity $\kappa\mu$ has the values 0·57, 0·49, 0·44 when $p = 6, 8, 10$ respectively. The above formulae hold only when the absorbing atom is very much heavier than the perturbing atom, which is approximately satisfied in the case of the broadening of the mercury line. Dividing $\Delta\nu_S$ by $\Delta\nu_L$, the very interesting result is obtained that

$$\frac{\Delta\nu_S}{\Delta\nu_L} = -\tfrac{1}{2}\tan\frac{\pi}{p-1} \qquad \dots\dots(118),$$

which provides a method of determining p from experimental values of $\Delta\nu_S$ and $\Delta\nu_L$. In Table XXIII the values of $\Delta\nu_S/\Delta\nu_L$ obtained by Füchtbauer, Joos and Dinkelacker for the broadening of the mercury resonance line by several foreign gases are given along with the values of p calculated therefrom. Recent values obtained by Margenau and Watson[54a] for the sodium D lines are also included.

TABLE XXIII

Absorbing atom	Foreign gas	$\dfrac{\Delta\nu_S}{\Delta\nu_L}$	p	ρ_0 in Å. from Eq. (117) and experiment	ρ_0 in Å. from Eq. (115) Theoretical
Hg	A	0·39	5·8	14·4	Between 11·7 and 18
,,	H_2	0·16	11	8·3	—
,,	N_2	0·45	5·3	12	—
,,	CO_2	0·244	8	18	—
,,	H_2O	0·22	8·6	13	—
Na	N_2	0·44	5·3	—	—
,,	A	0·34	6·4	—	Between 16·5 and 21·6
,,	H_2	0·23	8·3	—	—

With the aid of the known values of p and of the experimental measurements of $\Delta\nu_L$, ρ_0 can now be obtained from Eq. (117). These are listed in the fifth column of Table XXIII, and are seen to be larger than the values of σ_L obtained with the aid of the old Lorentz theory. It is interesting to note that argon, for which $p = 5\cdot8$, is the only gas to which the London theory ($p = 6$) can be applied. It is therefore worth while to compute the ρ_0 for argon from Weisskopf's value of ρ and to compare the resulting theoretical value with the value obtained from $\Delta\nu_L$. Since Weisskopf's value of ρ lies between 5·4 and 8·3 Å., and Lenz's ρ_0 is $1·50(2\pi)^{\frac{1}{p-1}}$ times as large, the theoretical value of ρ_0 comes out between 11·7 and 18 Å. in agreement with the value 14·4.

The frequency distribution of the Lorentz-broadened line was found by Lenz to be

$$J(\omega) = \frac{1 + \tanh\{\epsilon(\omega_0 + \Delta\omega_0 - \omega)\}\tau_0}{(\omega_0 + \Delta\omega_0 - \omega)^2 + \left(\dfrac{\Delta\omega_H}{2}\right)^2} \quad \ldots\ldots(119),$$

where $\epsilon = 0·31 - 0·011p$, $\omega_0 = 2\pi\nu_0$, $\Delta\omega_0 = 2\pi\Delta\nu_S$, $\Delta\omega_H = 2\pi\Delta\nu_L$, and $\tau_0 = \rho_0/c_0$. This formula represents a curve whose maximum is shifted by the amount $\Delta\omega_0 - \dfrac{\epsilon}{8}\Delta\omega_H{}^2 . \tau_0$. The second factor, however, is small in comparison with the first, and may be neglected when the foreign gas pressure is not too high. The experiments of Füchtbauer, Joos and Dinkelacker justify this, since the shift was found to be proportional to the foreign gas pressure. The quantity $\Delta\omega_H$ represents the half-breadth of the line. The formula must not be expected to hold at very great distances from the centre of the line.

As a last result, Lenz calculated the asymmetry of the line, by using, as a measure, the difference between the intensities at equal distances on both sides of the line, divided by the sum. He found

$$\frac{J_+ - J_-}{J_+ + J_-} = \tanh\{\epsilon(\omega_0 + \Delta\omega_0 - \omega)\tau_0\} \quad \ldots\ldots(120)$$

to be independent of the pressure of the foreign gas and to hold only near the centre of the line and for small foreign gas pressures (less than one atmosphere). Unfortunately, there are no experiments at hand with which to compare the above expression with any degree of accuracy.

It is remarkable that Lenz was able to obtain so many useful results on the basis of the simple law of interaction,

$$V'(r) - V(r) = -ar^{-p}.$$

This may mean that only the right-hand portions of the Franck-Condon curves are significant in treating Lorentz broadening. It would be an interesting problem, however, to represent analytically the complete Franck-Condon curves suggested by Kuhn and Oldenberg[40] from an analysis of the mercury-rare gas bands, and to follow through Lenz's theory with these expressions substituted for $V'(r) - V(r)$. It is to be expected, of course, that the mathematical difficulties would be very great.

5. HOLTSMARK BROADENING

The broadening of an absorption line that takes place when the pressure of the absorbing gas is increased (with no foreign gas present), which we have called Holtsmark broadening, was first treated by Holtsmark[30] on the basis of classical theory. He calculated the mean frequency shift due to interaction of similar oscillators and found it to vary as $\sqrt{Nf}$. Measurements of the Holtsmark broadening of the mercury and sodium resonance lines made by Trumpy[98] were apparently in agreement with Holtsmark's theory. There were, however, objections to the theory, and both Frenkel[21] and Mensing[57] attempted to handle the problem on the basis of quantum mechanics. Recently Schütz-Mensing[89] has pointed out that Holtsmark's original classical treatment was unjustified and, when carried out properly, gives rise to a line breadth that varies directly as Nf. In no case was it possible to calculate the actual line form.

Recently the problem has been attacked from an entirely different point of view by Weisskopf[105]. In the opinion of this author, Holtsmark broadening is to be regarded in the same light as Lorentz broadening, namely as a result of damping due to collisions among the absorbing atoms. On the basis of classical dispersion theory, the absorption coefficient of a gas when both natural damping and collision damping are present is found to be

$$k \propto \sqrt[4]{\frac{(b-\Delta)^2 + 1}{\Delta^2 + 1}} \sin \tfrac{1}{2} \arctan \frac{b}{\Delta(\Delta - b) + 1} \quad ...(121),$$

where $\qquad b = \dfrac{4\pi A}{\gamma}, \qquad A = \dfrac{1}{2}\dfrac{e^2 N f}{2\pi v_0 m}$ ......(122),

$$\gamma = \frac{1}{2\tau} + Z_H = \frac{1}{2\tau} + 4\sigma_H^2 N \sqrt{\frac{\pi RT}{M}} \qquad ......(123)$$

and $\qquad \Delta = \dfrac{2\pi(\nu - \nu_0)}{\gamma} + \dfrac{b}{3}$ ......(124).

In the above formula, N stands for the number of absorbing atoms per c.c., Z_H for the number of Holtsmark broadening collisions per sec. per c.c. per absorbing atom, and σ_H^2 for the effective cross-section associated with Holtsmark broadening. Assuming that the interaction between two absorbing atoms is equivalent to the interaction between two dipoles, Weisskopf was led to an approximate expression for σ_H^2, namely

$$\sigma_H^2 = \frac{2e^2 f}{4\bar{v}m \, 2\pi v_0} \qquad ......(125).$$

Introducing the above value of σ_H^2 into Eq. (123) and using Eqs. (122), b is found to be

$$b = \frac{\dfrac{e^2 N f}{v_0 m}}{\dfrac{1}{2\tau} + \dfrac{e^2 N f}{4v_0 m}} \qquad ......(126).$$

There are two important limiting cases. (1) When N is very small, b is very small, Δ is equal to $4\pi\tau(\nu - \nu_0)$, and Eq. (121) reduces to the classical dispersion formula

$$k \propto \frac{\dfrac{b}{2}}{1 + \Delta^2},$$

which, at the edges of the line, even in the presence of the Doppler effect, has been shown to reduce to

$$k \propto \frac{1}{\Delta^2}.$$

(2) When N is so large that collision damping far outweighs natural damping, $b = 4$, Δ is equal to $\dfrac{2\pi(\nu - \nu_0)}{Z_H} + \dfrac{4}{3}$, and the

absorption coefficient is given by the expression

$$k \propto \sqrt[4]{\frac{(4-\Delta)^2+1}{\Delta^2+1}} \sin \tfrac{1}{2} \arctan \frac{4}{\Delta(\Delta-4)+1} \dots (127),$$

which is represented graphically as a heavy curve in Fig. 42. Eq. (127) would still be expected to hold when the Doppler effect is present, provided the Holtsmark broadening was much larger than the Doppler broadening. The dotted curve in Fig. 42 is the simple Lorentz curve, and it is seen that the

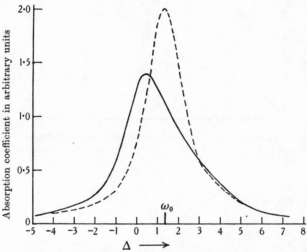

Fig. 42. Form of absorption line with large Holtsmark broadening.

heavy curve and the dotted curve agree at the edges of the line. We have then the result that, in the two extremes, vanishingly small absorbing gas pressure, and very high pressure, the frequency distribution at the edges of the line is identical, depending only on the damping. At intermediate values of N (when b lies between 0 and 4) the frequency distribution of the *edges of the line* will be given by

$$k \propto \frac{1}{\left[\dfrac{2\pi(\nu-\nu_0)}{\gamma}\right]^2} \qquad \dots\dots(128),$$

where
$$\gamma = \frac{1}{2\tau} + Z_H = \frac{1}{2\tau} + 4\sigma_H{}^2 N \sqrt{\frac{\pi RT}{M}} \qquad \dots\dots(129).$$

Now this is precisely the formula that has been used by Minkowski[60], Schütz[87] and Weingeroff[103] to calculate the damping from experiments performed on the edges of the sodium D lines (see Chap. III, § 5). If, therefore, Weisskopf's theory is correct, these experimentally determined values of

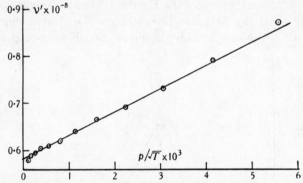

Fig. 43. Weingeroff's measurements of the damping of the NaD lines as a function of vapour pressure. (A smooth curve was drawn through the original experimental points.)

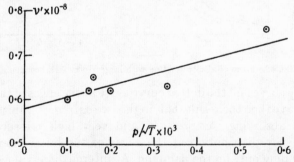

Fig. 44. Minkowski's measurements of the damping of the NaD lines as a function of vapour pressure.

the damping should vary linearly with $N\sqrt{T}$ or, in terms of the vapour pressure, with $p/\sqrt{T}$.

In Fig. 43, the damping ν' (which is exactly twice γ) is plotted against $p/\sqrt{T}$ for the experiments of Weingeroff, and in Fig. 44 for those of Minkowski. Both sets of experiments yield straight lines, but with different slopes. Since $\nu' = 2\gamma$,

$$\nu' = \frac{1}{\tau} + 2Z_H$$

$$= \frac{1}{\tau} + 8\sigma_H^2 N \sqrt{\frac{\pi RT}{M}}$$

$$= \frac{1}{\tau} + 8\sigma_H^2 \times 9\cdot74 \times 10^{18} \sqrt{\frac{\pi R}{M}} \cdot \frac{p}{\sqrt{T}},$$

and from Weingeroff's results:

$$\tau = 1\cdot7 \times 10^{-8}\,\text{sec.}, \quad \sigma_H = 14 \times 10^{-8}\,\text{cm.},$$

whereas from Minkowski's results:

$$\tau \doteq 1\cdot7 \times 10^{-8}\,\text{sec.}, \quad \sigma_H = 31 \times 10^{-8}\,\text{cm.}$$

These two values of σ_H are to be compared with Weisskopf's theoretical value 44×10^{-8} cm., obtained with the aid of Eq. (125).

Margenau [54] has given an interesting theory of Holtsmark broadening in terms of the screening effect upon a particular absorbing atom by other absorbing atoms that continually pass in and out of its line of sight. A wave-mechanical calculation yields the result that the half-breadth of an absorption line produced by such screening should be 0·445 divided by the mean time between collisions—a formula very much the same as Lorentz's formula. As far as the final formulas are concerned, Margenau and Weisskopf agree in their treatment of Holtsmark broadening. Their main difference of opinion lies in the fact that Weisskopf regards Holtsmark broadening as an example of Lorentz broadening, whereas Margenau considers that a *part* of Holtsmark broadening may be due to the screening effect of neighbouring atoms.

6. EARLY MEASUREMENTS OF THE QUENCHING OF RESONANCE RADIATION

6a. QUENCHING OF RESONANCE RADIATION BY FOREIGN GASES. It was first noticed by Wood [108] that the introduction of a small amount of air into a mercury resonance lamp reduced the intensity of the emitted resonance radiation. Further experiments on mercury, sodium and cadmium resonance radiation, to be described later, indicated that this is a very general phenomenon that takes place whenever the foreign gas atoms or molecules are capable of receiving some or all of the excitation energy of the excited atoms in the resonance lamp. In the current terminology, the resonance radiation whose intensity is reduced by the introduction of a foreign gas is said to be "quenched", and a collision between an excited atom and a foreign gas molecule in which some or all of the excitation energy of the excited atom is given over to the foreign gas molecule, thereby preventing the excited atom from

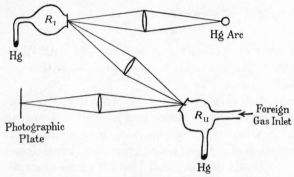

Fig. 45. Stuart's apparatus for studying the quenching of Hg resonance radiation.

radiating, is said to be a "quenching collision". The "quenching", Q, is defined as

$$Q = \frac{\text{Intensity of resonance radiation with foreign gas}}{\text{Intensity of resonance radiation without foreign gas}},$$

and the results of a quenching experiment are expressed by a "quenching curve", in which Q is plotted as a function of the foreign gas pressure.

6b. EXPERIMENTS OF STUART WITH MERCURY. The most extensive early investigation of the quenching of mercury resonance radiation was made by Stuart[93], who used the apparatus shown in Fig. 45. The main resonance lamp R_{II} into which the foreign gases were introduced was excited by the very narrow line emitted by the resonance lamp R_{I}, and the quenching of the resonance radiation from R_{II} was measured with a number of foreign gases. The experimental results are shown in Fig. 46 in the form of quenching curves. The following features of Stuart's experiments are to be emphasized: (1) The mercury vapour pressure corresponded to room temperature, at which there was considerable diffusion of imprisoned resonance radiation. (2) The main resonance lamp R_{II} was excited by a very narrow line. (3) In the case of the inert gases, He, A and N_2, appreciable quenching occurred only at high foreign

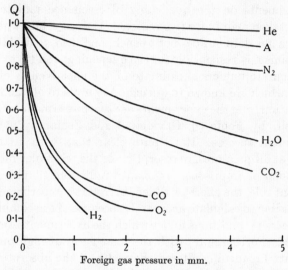

Fig. 46. Stuart's quenching curves for Hg resonance radiation.

gas pressures, from 10 to 200 mm. (4) Even at high pressures, the quenching of the inert gases was very much smaller than that of H_2, O_2, CO, CO_2 and H_2O.

6c. EXPERIMENTS WITH SODIUM AND CADMIUM. With apparatus similar to that of Stuart, Mannkopff[52] measured the quenching of sodium resonance radiation by H_2, N_2 and a mixture, Ne–He. He found that both N_2 and H_2 were very effective in quenching, and that Ne–He was very ineffective, even at high pressures. The sodium vapour pressure in these experiments was high enough to give rise to considerable reabsorption of the resonance radiation on its way out of the lamp.

The quenching of cadmium resonance radiation 3261 was studied qualitatively by Bates[5] and by Bender[8], who showed that H_2 was very effective in quenching cadmium resonance radiation. Further experiments by Bender indicated that N_2 and CO also quench, but not as effectively as H_2.

6*d*. DIFFICULTY OF INTERPRETING EARLY EXPERIMENTS. Experiments on the quenching of resonance radiation are undertaken mainly for two reasons: (1) to ascertain whether a foreign gas does or does not quench, and if so, to decide what mechanism is responsible for the quenching; and (2) to obtain an accurate numerical estimate of the effectiveness of those gases which are known to quench. Attempts to obtain this information from the experiments of Stuart were made by Stuart himself, by Foote[19], Gaviola[24] and Zemansky[112], with indifferent success. It is quite clear that, with all foreign gases at all pressures, reabsorption of the resonance radiation (diffusion of imprisoned resonance radiation) played an important rôle, the effect of which, in Stuart's experiments, it is impossible to calculate accurately, because of the complicated geometrical conditions under which the experiments were performed. Furthermore, with the inert gases at high pressure, Lorentz broadening altered the width of the absorption line relative to the width of the exciting line to such an extent that it is doubtful whether the so-called quenching curves obtained with the inert gases can be regarded as being due to quenching at all. Mannkopff's experiments on sodium resonance radiation also are difficult to interpret, not only because of reabsorption of resonance radiation but also because of a reaction which appears to take place between normal sodium atoms and some foreign gases, notably nitrogen and hydrogen, with the result that the sodium vapour pressure, and consequently the rate of formation of excited atoms, is reduced, causing a reduction in the intensity of the emitted resonance radiation that is not to be confused with the phenomenon of quenching.

It is rather important to emphasize that a quenching curve,

that is, an experimental curve of Q against foreign gas pressure, by itself, without further details as to line breadths, vapour pressure, geometry of apparatus, etc., can give no information *of an absolute nature* whatever, and indeed in some cases is not convincing evidence that quenching takes place at all. To read from a quenching curve the half-value pressure (the foreign gas pressure at which $Q = 0.5$), and to say that, at this pressure, the time between collisions is equal to the life-time of the excited atoms, may lead to errors of several hundred per cent. The most that can be inferred from a *series* of quenching curves, all taken under the same conditions but with different foreign gases, is the relative quenching ability (if it *is* quenching) of the various gases; and, as an approximate method of describing the relative effectiveness of various gases, the half-value pressure may be used. To obtain accurate information of an absolute character about quenching collisions between excited atoms and foreign gas molecules, methods must be developed to enable one to take into account the effect of radiation diffusion and of Lorentz broadening, or, better still, the experiments should be performed under conditions in which these effects are absent. To see how this is done it is necessary to consider these two effects at a little greater length.

7. THEORY OF THE QUENCHING CURVE FROM AN IDEAL RESONANCE LAMP

7 *a*. THE STERN-VOLMER FORMULA. Let us suppose that a beam of radiation is incident upon an ideal resonance lamp such as that depicted in Fig. 24, and let us assume that the following conditions are fulfilled:

(1) The absorbing gas in the resonance lamp is at such a low pressure that only primary resonance radiation is emitted which is not further absorbed on its way out.

(2) There is a foreign gas present at such a low pressure that Lorentz broadening of the absorption line is negligibly small.

If n represents the number of excited atoms per c.c. in the emitting layer of the resonance lamp, τ the lifetime of the excited atoms, and Z_Q the number of times per sec. that an excited atom gives up its excitation energy upon collision to one c.c. of molecules of foreign gas (number of quenching collisions per sec. per c.c. per excited atom), then, in the steady state,

Rate at which excited atoms $\Big\}$ = $\Big\{$ Rate at which excited atoms are being destroyed $\Big\}$ = $\Big\{$ are being formed

or
$$\frac{n}{\tau} + Z_Q n = E,$$

and
$$n = \frac{E}{\dfrac{1}{\tau} + Z_Q},$$

where E, under the conditions imposed above, is independent of the foreign gas pressure, and remains constant so long as the absorbing gas pressure and the intensity of the exciting light remain constant.

The radiation emitted by the resonance lamp is a constant fraction (depending on geometry), say ϵ, of the total energy emitted by the n excited atoms, thus

Emitted radiation *with* foreign gas $= \epsilon \dfrac{n}{\tau} h\nu$

$$= \epsilon h\nu E \frac{1}{1 + \tau Z_Q}.$$

Now, the emitted radiation *without* foreign gas $= \epsilon h\nu E$, whence

$$Q = \frac{1}{1 + \tau Z_Q} \qquad \qquad \ldots\ldots(130).$$

This formula was first obtained by Stern and Volmer[92]. Since Z_Q varies linearly with the foreign gas pressure, a quenching curve obtained in an experiment performed under these conditions should follow a simple curve of the type

$$Q = \frac{1}{1 + \text{const.}\, p},$$

or, plotting $1/Q$ against p, a straight line should result. It would seem, therefore, that the criterion for applying the Stern-Volmer formula to an experimental quenching curve would be to see whether the experimental values of $1/Q$ vary linearly with p. It appears, however, that this is not a very sensitive criterion. For example, Stuart's quenching curve obtained with H_2 obeys this formula, whereas the other curves do not, and yet the formula is equally inapplicable to all the quenching curves. Similarly, both Mannkopff's and von Hamos' experimental values [26] for the quenching of sodium resonance radiation by N_2 obey the formula, although Mannkopff's experiments were performed under conditions which definitely preclude its use. The fact of the matter is, that almost any small portion of a descending curve can be fitted with some degree of accuracy to a formula of the Stern-Volmer type. The only way to tell whether the use of the formula is justified is to test whether the conditions that are assumed in the derivation are satisfied or not. For example, in working with sodium resonance radiation, it is easy to see whether primary resonance radiation alone is excited, or whether radiation diffusion is present, by merely noticing whether resonance radiation is coming only from the direct path of the exciting light or from the resonance lamp as a whole. In the case of ultra-violet resonance radiation, the easiest procedure is to obtain several quenching curves with the same foreign gas but at different absorbing gas pressures. If the *same* quenching curve is obtained at various low absorbing gas pressures, then the region in which the Stern-Volmer formula is applicable has been attained. If not, the absorbing gas pressure must be reduced until the quenching curve becomes constant.

The only quenching experiment, involving the excitation of normal atoms in a resonance lamp by the resonance line, which seems to warrant the use of the Stern-Volmer formula, is that of von Hamos on the quenching of sodium resonance radiation by N_2. The value of the effective quenching cross-section σ_Q^2 obtained from von Hamos' experiments is listed in a table at the end of § 8.

$7b$. Effect of Lorentz Broadening on Quenching. Even when the absorbing gas pressure is low, the Stern-Volmer formula may be inapplicable. This is the case when the foreign gas pressure is high enough to produce Lorentz broadening of the absorption line. To understand how this affects the quenching curve, let us consider two extreme cases: (1) the exciting line is a broad, self-reversed line, such as that emitted by a hot arc; and (2) the exciting line is a narrow, unreversed line such as that emitted by a resonance lamp. These two cases are depicted in Fig. 47. The figures show the relation between the exciting line

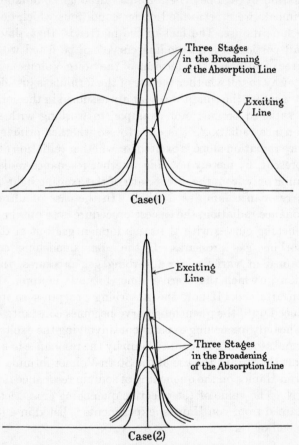

Fig. 47. Lorentz broadening in a resonance lamp.

(which remains constant) and the absorption line of the gas in the resonance lamp as it is broadened by increasing the foreign gas pressure. The area on the graph, which the exciting line and the absorption line have in common, is an indication of the energy absorbed in the resonance lamp, and therefore of the rate at which excited atoms are forming. It is evident from the figure illustrating case (1) that, when the exciting line is wide, this area increases as the absorption line gets broader. In this case, therefore, if the foreign gas does not quench at all, the quenching curve should *rise*, and if quenching does take place, it would be partly or completely offset by the increased absorption. In case (2), it is evident that the net effect of Lorentz broadening of the absorption line is to cause the emitted radiation to decrease *whether real quenching is present or not*.

The experiments of Bates[6] and von Hamos[26] are in support of these conclusions. Bates obtained the quenching curve of mercury resonance radiation for the foreign gas methane. For small values of the methane pressure (0 to 10 mm.), the curve descended slightly, due to a small amount of true quenching. When the methane pressure was increased (10 to 200 mm.), the quenching curve showed a marked rise. Since the exciting light was obtained from an arc, the conditions under which Bates worked were equivalent to case (1).

Von Hamos studied the emission of sodium resonance radiation under conditions in which the exciting line could be made broad or narrow by running the arc hot or cold, and in which the absorption line could be either broadened by a foreign gas, or shifted with respect to the exciting line by a magnetic field. His results confirm in every detail the conclusions stated above, and also indicate that the inert gases do not quench sodium resonance radiation at all. It seems to be quite certain from these considerations that the quenching of mercury resonance radiation by the inert gases observed by Stuart, and the quenching of sodium resonance radiation by the mixture Ne–He observed by Mannkopff, are not true quenching at all— but are *due entirely to Lorentz broadening*.

In view of the difficulty of interpreting Stuart's quenching curves, experiments were performed by Zemansky under conditions in which Lorentz broadening was absent but radiation diffusion played the main rôle. Before these experiments can be understood, it is necessary to give a brief account of Milne's theory of radiation diffusion.

8. RADIATION DIFFUSION AND QUENCHING

8*a*. MILNE'S THEORY. Consider a mass of gas, enclosed between the planes $x = 0$ and $x = l$, exposed to isotropic mono-chromatic radiation at the face $x = 0$, which is capable of raising atoms from the normal state 1 to the excited resonance state 2. Suppose at any moment that there are n_1 normal atoms per c.c. capable of absorbing this radiation and n_2 excited atoms per c.c. capable of emitting this radiation. Then it has been shown by Milne[59], on the basis of Einstein's radiation theory, that n_2 at any point is given by

$$\frac{\partial^2}{\partial x^2}\left(n_2 + \tau\frac{\partial n_2}{\partial t}\right) = 4k^2\tau\frac{\partial n_2}{\partial t} \qquad \ldots\ldots(131),$$

where τ is the lifetime of the excited atoms, and k is the absorption coefficient of the gas in cm.$^{-1}$ (k contains n_1) for the radiation in question. This equation holds for all values of n_2 and n_1 provided $n_2 \ll n_1$, which is undoubtedly the case for light intensities employed in the laboratory. Dividing the radiation into two parts, in the manner of Schuster[86], Milne showed that the net forward flux of radiation at any point is given by

$$\pi I_+ = \frac{\pi\sigma'}{n_1}\left[\left(n_2 + \tau\frac{\partial n_2}{\partial t}\right) - \frac{1}{2k}\frac{\partial}{\partial x}\left(n_2 + \tau\frac{\partial n_2}{\partial t}\right)\right] \ldots\ldots(132),$$

and the net backward flux by

$$\pi I_- = \frac{\pi\sigma'}{n_1}\left[\left(n_2 + \tau\frac{\partial n_2}{\partial t}\right) + \frac{1}{2k}\frac{\partial}{\partial x}\left(n_2 + \tau\frac{\partial n_2}{\partial t}\right)\right] \ldots\ldots(133),$$

where

$$\sigma' = \frac{g_1}{g_2}\cdot\frac{2h\nu^3}{c^2} \qquad \ldots\ldots(134),$$

and g_1 and g_2 are the statistical weights of the normal and the excited states respectively.

The $\partial n_2/\partial t$ that appears in these expressions represents the resultant rate of formation of excited atoms under the influence of the three processes: absorption of radiation, spontaneous emission and stimulated emission. If there should be any other rate of formation of excited atoms, say R, then $\partial n_2/\partial t$ must be replaced by $\partial n_2/\partial t - R$. It must be emphasized that the above theory is good only for the one-dimensional flow of radiation, and that the motions of individual atoms and any radiation frequency changes that accompany such motions have not been taken into account.

8b. USE OF MILNE'S THEORY TO STUDY QUENCHING. In order to apply Milne's theory to the actual conditions of a quenching experiment, consider the arrangement depicted in Fig. 48. The absorption cell containing the absorbing gas and a foreign gas is an experimental approximation to an infinite

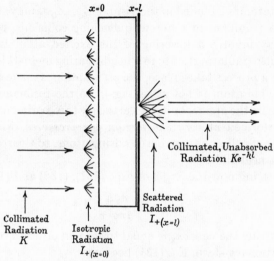

Fig. 48. Conditions postulated by Milne.

slab. Incident on the face $x = 0$ is the isotropic radiation postulated by Milne and, in addition, a collimated beam of intensity K. The effect of the collimated beam is to provide a further rate of formation of excited atoms equal to

$$B_{1\to 2} n_1 \frac{K}{4\pi} e^{-kx}.$$

By virtue of Einstein's relation,

$$B_{1\to 2} = \frac{g_2}{g_1} \cdot \frac{c^2}{2h\nu^3} \cdot \frac{1}{\tau} = \frac{1}{\sigma'\tau},$$

we can write

$$\left.\begin{array}{l}\text{Rate of formation of excited atoms} \\ \text{due to collimated beam}\end{array}\right\} = \frac{n_1 K}{4\pi\sigma'\tau} \cdot e^{-kx} \quad\ldots\ldots(135).$$

The effect of the presence of the foreign gas is to provide a further rate of destruction of excited atoms equal to

$$\left.\begin{array}{l}\text{Rate of destruction of excited atoms}\\ \text{due to quenching collisions}\end{array}\right\} = Z_Q n_2 \dots\dots(136),$$

where Z_Q depends on the foreign gas pressure according to Eq. (91). It must be emphasized that *all* collision processes between an excited atom and a foreign gas molecule, at the conclusion of which the excited atom is no longer in exactly the same state, are included in the symbol Z_Q. $Z_Q n_2$ may therefore include a number of different quenching collisions, of which one may involve a lowering of the excited atom from the resonance (radiating) state to a neighbouring metastable level. If such a process takes place, the assumption is made that the metastable atom is not raised again to the resonance state. This limits the applicability of the theory to experiments which are performed at *very low foreign gas pressures*, where the number of collisions capable of raising atoms to the resonance state is negligible.

Replacing now the $\partial n_2/\partial t$ of Eqs. (131), (132) and (133) by

$$\frac{\partial n_2}{\partial t} + Z_Q n_2 - \frac{n_1 K}{4\pi\sigma'\tau} \cdot e^{-kx},$$

and putting the new $\partial n_2/\partial t$ equal to zero, in order to represent the stationary state, Eq. (131) becomes

$$\frac{\partial^2 n_2}{\partial x^2} = 4k^2 \frac{\tau Z_Q}{1 + \tau Z_Q} n_2 - 3 \frac{k^2 n_1 K}{4\pi\sigma'(1 + \tau Z_Q)} \cdot e^{-kx} \dots\dots(137),$$

and Eqs. (132) and (133) become

$$\pi I_+ = \frac{\pi\sigma'}{n_1}(1 + \tau Z_Q)\left(n_2 - \frac{1}{2k}\frac{\partial n_2}{\partial x}\right) - \frac{3}{2}\frac{K}{4}e^{-kx} \dots(138),$$

$$\pi I_- = \frac{\pi\sigma'}{n_1}(1 + \tau Z_Q)\left(n_2 + \frac{1}{2k}\frac{\partial n_2}{\partial x}\right) - \frac{1}{2}\frac{K}{4}e^{-kx} \dots(139).$$

If now we do away entirely with the isotropic radiation, and keep only the collimated beam, we have the boundary conditions that

$$\text{when} \quad x = 0, \quad I_+ = 0;$$
$$\text{when} \quad x = l, \quad I_- = 0.$$

The details of the solution of Eqs. (137), (138) and (139), subject to the boundary conditions above, are given in a paper by Zémansky[115], where it is shown that the scattered radiation emerging from unit area of the face $x = l$ is given by

$$\pi I_+ (x = l) = K G (kl, \tau Z_Q) \qquad \ldots\ldots(140),$$

where $G (kl, \tau Z_Q)$

$$= \frac{1}{2(1 - 3\tau Z_Q)} \left[\frac{6\sqrt{\tau Z_Q (1 + \tau Z_Q)} + e^{-kl} \sinh 2kl \sqrt{\dfrac{\tau Z_Q}{1 + \tau Z_Q}}}{\sinh \left(2kl \sqrt{\dfrac{\tau Z_Q}{1 + \tau Z_Q}} + 2 \sinh^{-1} \sqrt{\tau Z_Q} \right)} - 3e^{-kl} \right]$$

$$\ldots\ldots(141).$$

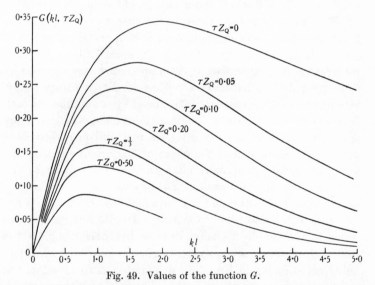

Fig. 49. Values of the function G.

It is clear that the function G depends only on the two quantities kl (called, in astrophysics, the "opacity"), and τZ_Q, the number of quenching collisions per lifetime. In a table in the Appendix, values of G are given for many values of kl and τZ_Q, and the function G is plotted in Fig. 49. It can be seen from the figure and can also be shown from Eq. (141) that, as $kl \to 0$,

$$G \to \frac{1}{1 + \tau Z_Q} \cdot \frac{kl}{2},$$

whence the quenching Q approaches

$$Q \to \frac{1}{1 + \tau Z_Q},$$

which is the Stern-Volmer formula of Eq. (130).

Since the above calculation concerns an infinitesimally narrow range of frequencies which are not altered upon repeated absorptions and re-emissions, it is necessary to develop methods for handling the actual situation in which the exciting light is a fairly broad spectral line, and in which the absorbing gas has an absorption line with a Doppler distribution. It if be assumed that each narrow frequency band $d\nu$, present in the exciting light, diffuses without change of frequency according to Eq. (140) with the appropriate $k_\nu l$, then a knowledge of the frequency distribution of both the exciting and the absorption lines will enable one to integrate graphically to obtain the resultant emerging radiation. This procedure was adopted by Zemansky with some success, but has the disadvantage that it is tedious, and that the error introduced by assuming that each frequency band diffuses as a unit without change of frequency is difficult to estimate. It is far simpler and probably more accurate to assume that the effect of the broad exciting line and the various frequency changes that take place as the radiation diffuses is *the same as that which would be produced by an infinitesimal frequency band for which the absorbing gas has an "equivalent absorption coefficient"*. In other words, we can describe the whole diffusion process by attributing to the diffusing radiation an equivalent opacity, say $\bar{k}l$, which, when substituted for kl, in Eq. (140) will enable us to calculate the intensity of the emerging radiation. The success of this method, of course, depends entirely upon the degree of accuracy with which the equivalent opacity can be calculated. Samson[84] has given a method of calculating an equivalent opacity which is reliable at *low* absorbing gas pressures at which most experiments are performed.

8c. EQUIVALENT OPACITY AT LOW PRESSURE. Samson assumed that the scattered or diffused radiation had a frequency distribution determined only by the Doppler effect and independent of the breadth of the exciting line. He defined an equivalent absorption coefficient as that value of kl, say $\bar{k}l$, which a gas would have to possess for an infinitesimal frequency band in order that this infinitesimal band would be transmitted to the same extent that the actual Doppler radiation is transmitted. It has already been shown in Chap. III, §4d, that the transmission of a Doppler line is given by [see Eq. (61)]

$$\frac{\int_{-\infty}^{\infty} e^{-\omega^2} . e^{-k_0 l e^{-\omega^2}} d\omega}{\int_{-\infty}^{\infty} e^{-\omega^2} d\omega},$$

where, for a simple line [see Eq. (35)],

$$k_0 = \frac{2}{\Delta\nu_D} \sqrt{\frac{\ln 2}{\pi}} . \frac{\lambda_0^2 g_2}{8\pi\tau g_1} . N.$$

The transmission, however, of an infinitesimal frequency band for which the gas has an opacity $\bar{k}l$ is $e^{-\bar{k}l}$, whence $\bar{k}l$ can be calculated from the formula

$$e^{-\bar{k}l} = \frac{\int_{-\infty}^{\infty} e^{-\omega^2} . e^{-k_0 l e^{-\omega^2}} d\omega}{\int_{-\infty}^{\infty} e^{-\omega^2} d\omega} \qquad \ldots\ldots(142),$$

once $k_0 l$ is known. There will be found in the Appendix a table of values of Samson's $\bar{k}l$ for a number of values of $k_0 l$.

8d. Derivation of a Theoretical Quenching Curve. Once the equivalent opacity has been calculated, the scattered radiation emerging from the face $x = l$ can be obtained for a number of values of τZ_Q from the curves of Fig. 49. Dividing by the result when $\tau Z_Q = 0$, the quenching Q is obtained as a function of τZ_Q. A number of such theoretical quenching curves are shown in Fig. 50 along with the Stern-Volmer curve, which is valid only when $\bar{k}l$ is vanishingly small. From the correct theoretical quenching curve the value of τZ_Q may be read off corresponding to any experimentally observed value of Q, and therefore, an experimental quenching curve of Q against foreign gas pressure can be converted into a curve of

τZ_Q against foreign gas pressure which, by virtue of the gas-kinetic expression for Z_Q, should be a straight line whose slope contains $\sigma_Q{}^2$, the quenching cross-section.

8e. EXPERIMENTAL DETERMINATIONS OF QUENCHING CROSS-SECTIONS. Zemansky[115] and Bates[6, 7] used the apparatus depicted in Fig. 51. Resonance radiation from a mercury resonance lamp R was passed through a thin quartz absorption cell containing mercury vapour and a foreign gas

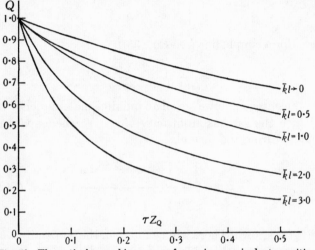

Fig. 50. Theoretical quenching curves for various equivalent opacities.

at some known pressure. The difference between the readings of the photoelectric cell, P, in positions (1) and (2) gave the intensity of the scattered radiation, since, in position (1) the photoelectric cell received only the transmitted collimated beam, whereas in position (2) both the transmitted collimated beam and the scattered radiation were received. The incident radiation was measured often (through the cellophane window C, which cut it down to an easily measurable value), in order to take account of any variation in intensity of the radiation from R. The ratio of the intensity of the scattered radiation with a foreign gas to that without a foreign gas gave the quenching Q, which was measured for many foreign gases.

In Zemansky's experiments, the mercury vapour pressure in the absorption cell corresponded to a temperature of 20·0° C. throughout. Assuming the mercury resonance line to consist of five equal hyperfine-structure components, and taking τ to be $1\cdot08 \times 10^{-7}$ sec., $k_0 l$ is found to be, from Eq. (35),

$$k_0 l = 1\cdot31 \times 10^{-13} N l.$$

At 20° C., $N = 4\cdot19 \times 10^{13}$, and l was 0·792 cm., whence $k_0 l$ was found to be 4·35. Substituting this value of $k_0 l$ in Samson's

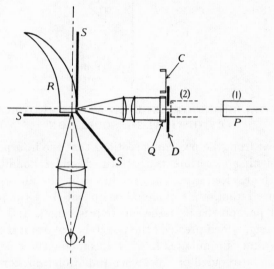

Fig. 51. Quenching apparatus satisfying the provisions of Milne's theory.

formula, Eq. (142), $\bar{k}l$ was calculated to be 2·24. From the curves in Fig. 49, the theoretical quenching curve appropriate to these experiments was plotted. From the theoretical quenching curve, and experimental measurements of Q at various foreign gas pressures with several foreign gases, curves of τZ_Q against p were obtained, and, from the slopes of these lines and the gas-kinetic expression for Z_Q, the value of $\sigma_Q{}^2$ appropriate to each foreign gas was finally obtained. All such values are given in Table XXIV along with values obtained by Bates. All the foreign gases were studied within a pressure range in which Lorentz broadening was entirely negligible. In this pressure range no appreciable quenching was observed for helium or for argon, proving that Stuart's quenching curves for these gases are due entirely to Lorentz broadening.

TABLE XXIV

Foreign gas molecule	σ_v^2 cm.$^2 \times 10^{16}$	Foreign gas molecule	σ_q^2 cm.$^2 \times 10^{16}$
CH_4	0·0852	O_2	19·9
H_2O	1·43	H_2	8·60
NH_3	4·20	CO_2	3·54
NO	35·3	C_2H_6	5·94
CO	5·82	C_3H_8	2·32
N_2	0·274	C_4H_{10}	5·88
		C_6H_6	59·9
		Other hydro-carbons	Very large

9. COLLISIONS OF EXCITED ATOMS PRODUCED BY OPTICAL DISSOCIATION

It is clear from the preceding sections that the interpretation of quenching experiments, performed under conditions in which the excited atoms are produced by the absorption of resonance radiation, is rendered complicated by the presence of such phenomena as radiation imprisonment and Lorentz broadening, which preclude the possibility of using the simple Stern-Volmer formula. It has been seen how these complications can be avoided or how, when radiation imprisonment is present, it may be taken into account by Milne's theory. Since, however, this last method is very tedious, it is therefore fortunate that another method is at hand which is not only simpler from an experimental and a theoretical point of view, but is also more powerful as a tool for studying quenching.

9a. THE OPTICAL DISSOCIATION OF NaI. This method depends upon a very important process first observed by Terenin[95], namely, the dissociation of the NaI molecule into an *excited* sodium atom and a normal iodine atom by light of wave-length 2430 or less. The process may be represented by the equation:

$$NaI + h\nu\,(\lambda = 1900 \text{ to } 2430) = Na\,(3\,^2P) + I.$$

Since the dissociation potential of NaI into an excited sodium atom and a normal iodine atom is 5·078 volts, corresponding to $\lambda = 2430$ Å., it is clear that, when the dissociation is accomplished by means of light of wave-length shorter than 2430, the excess energy is transformed into relative kinetic energy of the resulting atoms, as was shown by Hogness and Franck [29]. The excited atoms formed by the dissociation emit the D lines, whose intensity in the presence of a foreign gas is diminished, enabling one to measure the quenching under particularly advantageous conditions, namely, (1) at any moment the concentration of normal sodium atoms is very small, hence very little of the D light is absorbed by sodium atoms on its way out of the vessel, and the Stern-Volmer formula can be used with confidence; (2) the velocity of the excited sodium atom can be varied at will by performing the dissociation of NaI with various wave-lengths, and therefore the dependence of the effective quenching cross-section on the velocity can be studied; (3) foreign gases may be used which react chemically with normal sodium atoms.

9b. EXPERIMENTAL RESULTS. Three different molecules have been used for the production of excited atoms. Winans [107], Terenin and Prileshajewa [96], and Kisilbasch, Kondratjew and Leipunsky [36] used NaI, and Winans, in another experiment, used NaBr to produce excited sodium atoms, and Prileshajewa [79, 80] used TlI to produce excited thallium atoms. The apparatus in all cases was substantially the same, so that a description of Terenin and Prileshajewa's procedure with NaI will suffice. Solid NaI was warmed to about 550° C. in a quartz vessel until the vapour pressure was about 0·015 mm. Exciting light from a spark was focused on a warmer part of the vessel and the D radiation, emitted perpendicular to the exciting light, was measured as a function of the foreign gas pressure. A visual photometric method was used in all cases, a second vessel containing NaI without a foreign gas being used as a comparison standard by Winans, and another source of yellow light being employed for this purpose by Terenin and Prileshajewa. Winans used three different exciting wave-lengths, whereas Terenin and Prileshajewa used eight different wave-lengths. Corrections were made by Terenin and Prileshajewa for the absorption of both the exciting light and the emitted D light by the foreign gas.

In Fig. 52 are shown some typical quenching curves in which the exciting source was a cadmium spark and the dissociated molecule was NaI. The fact that these curves obey the Stern-Volmer formula is indicated in Fig. 53, where $1/Q$ is plotted against foreign gas pressure and a straight line is

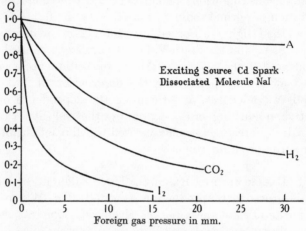

Fig. 52. Quenching of Na resonance radiation by foreign gases.

obtained. The effect of the wave-length of the exciting source upon the quenching is shown in Fig. 54, where the quenching of I_2 is shown for three different exciting sources.

9c. EVALUATION OF EFFECTIVE CROSS-SECTIONS. It has already been mentioned that, if the dissociation of the NaI molecule is accomplished by means of light of wave-length less than 2430, the excess energy is transformed into relative kinetic energy of the resulting atoms. In calculating the collision rate of excited atoms it is incorrect, therefore, to attribute to them the usual kinetic energy of thermal motion. It was shown by Terenin and Prileshajewa that, in comparison to the speed of the excited sodium atom after dissociation, the molecules of NaI could be regarded, with negligible error, to be stationary,

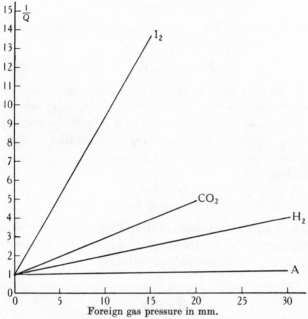

Fig. 53. Applicability of Stern-Volmer formula.

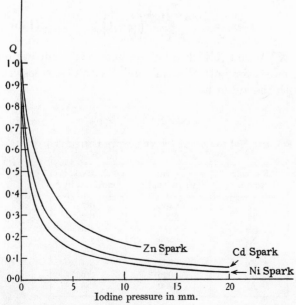

Fig. 54. Dependence of quenching on velocity of excited Na atom.

in which case the number of quenching collisions per excited sodium atom per sec. per c.c. is given by

$$Z_Q = \pi \sigma_Q{}^2 n_2 v_1 \qquad \ldots\ldots(143),$$

where n_2 is the number of foreign gas molecules per c.c. and v_1 is the velocity with which the excited sodium atom escapes from the NaI molecule upon dissociation. Since the velocity of escape is uniformly distributed as regards direction in space, as was shown by Mitchell[62], it follows from elementary mechanics that

$$v_1{}^2 = \frac{m_2}{m_1} \cdot \frac{2w}{m_1 + m_2} \qquad \ldots\ldots(144),$$

where m_1 and m_2 are the masses of the sodium atom and iodine atom respectively, and w is the excess energy of the dissociating quantum. Using the Stern-Volmer formula:

$$Q = \frac{1}{1 + \tau Z_Q},$$

with $\tau = 1 \cdot 6 \times 10^{-8}$ sec., and denoting the foreign gas pressure in mm. by p, we get finally

$$\sigma_Q{}^2 = \frac{1 \cdot 83 \times 10^{-9}}{p v_1} \left(\frac{1}{Q} - 1 \right) \qquad \ldots\ldots(145).$$

In Tables XXV and XXVI are given values of v_1 for various exciting sources for both the NaI and NaBr molecules, calculated with the aid of Eq. (144).

TABLE XXV

FOR THE NaI MOLECULE (WAVE-LENGTH LIMIT 2430)

Exciting source	Mean wave-length	Velocity v_1 in cm./sec. $\times 10^{-5}$
Fe	2400	0·7
Tl	2380	0·7
Sb	2311	1·3
Ni	2300	1·4
Cd	2232	1·7
Zn	2082	2·4
Mg	2026	2·6
Al	1990	2·8

TABLE XXVI

FOR THE NaBr MOLECULE (WAVE-LENGTH LIMIT 2144)

Exciting source	Mean wave-length	Velocity v_1 in cm./sec. $\times 10^{-5}$
Cd	2232	0·4
Zn	2082	1·2
Al	1990	1·8

In Table XXVII, the quenching cross-sections, calculated with the aid of Eq. (145), are listed along with the values obtained by Mannkopff and von Hamos. Comparing Mannkopff's

TABLE XXVII

Optically dis-sociated molecule	Foreign gas	Excitation	Velocity of excited Na atom in cm./sec. $\times 10^{-5}$	σ_Q^2 cm.$^2 \times 10^{16}$	Author
—	H_2	Resonance	~0·7	17 (corrected about 8)	Mannkopff
NaI	,,	Cd	1·7	6·4	Winans
,,	,,	Zn	2·4	5·7	,,
—	N_2	Resonance	~0·7	61	Mannkopff
—	,,	,,	~0·7	29	von Hamos
NaI	,,	Zn	2·4	6·08	Kisilbasch, Kondratjew and Leipunsky
,,	,,	Al	2·8	10·6	,,
,,	,,	Al	2·8	9·6	Winans
,,	CO	Zn	2·4	4·04	Kisilbasch, Kondratjew and Leipunsky
,,	CO_2	Cd	1·7	16·9	Winans
,,	I	Fe	0·7	38·2	Terenin and Prileshajewa
,,	,,	Ni	1·4	60·5	,,
,,	,,	Zn	2·4	41·4	,,
,,	I_2	Fe	0·7	239	,,
,,	,,	Tl	0·7	191	,,
,,	,,	Sb	1·3	127	,,
,,	,,	Ni	1·4	153	,,
,,	,,	Cd	1·7	89·2	,,
,,	,,	Zn	2·4	38·2	,,
,,	,,	Mg	2·6	47·8	,,
,,	,,	Al	2·8	54·1	,,
NaBr	Br_2	Cd	0·4	366	Winans
,,	,,	Zn	1·2	124	,,
,,	,,	Al	1·8	102	,,

value of $\sigma_Q{}^2$ for nitrogen with that of von Hamos (which was obtained under more advantageous experimental conditions), it is seen to be approximately twice as large. Assuming then that Mannkopff's value for H_2 is also about twice as large as the correct value, we can estimate a corrected value for H_2. This is given in brackets. The dependence of $\sigma_Q{}^2$ on the velocity of

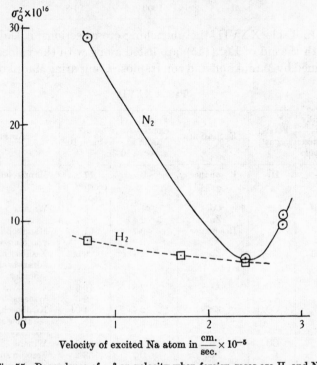

Fig. 55. Dependence of $\sigma_Q{}^2$ on velocity when foreign gases are H_2 and N_2.

the excited sodium atom is shown graphically in Figs. 55 and 56.

In the experiments of Prileshajewa with excited thallium atoms, the TlI molecule was dissociated by light of wavelength less than 2100 into a thallium atom in the $7\,{}^2S_{1/2}$ state and a normal iodine atom, according to the equation

$$h\nu\,[\lambda < 2100] + \text{TlI} = \text{Tl}\,(7\,{}^2S_{1/2}) + \text{I}.$$

A thallium atom in the $7\,{}^2S_{1/2}$ state may return spontaneously

either to the normal $6\,^2P_{1/2}$ state, emitting the line 3776, or to
the metastable state $6\,^2P_{3/2}$, emitting the green line 5350. The
reduction in intensity of either of these lines, as the pressure
of the foreign gas is increased, may be used to measure the
quenching of the $7\,^2S_{1/2}$ state. Prileshajewa measured the

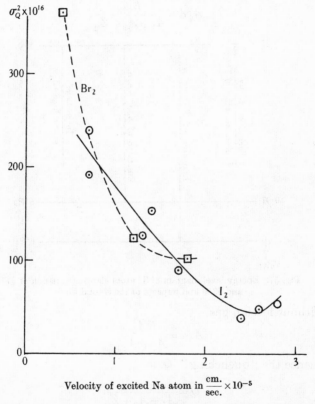

Fig. 56. Dependence of $\sigma_u{}^2$ on velocity when foreign gases are I_2 and Br_2.

quenching of the green line for three exciting wave-lengths and
with three foreign gases: I_2, I and TlI.

The derivation of the quenching formula appropriate to this
case is accomplished by appeal to Fig. 57, showing the various
processes that take place. Calling the rate of formation of
$7\,^2S_{1/2}$ atoms E, and denoting the Einstein A coefficients of the

lines 3776 and 5350 by A_1 and A_2 respectively, the number of excited atoms per c.c., n, is given by

$$E = A_1 n + A_2 n + Z_Q n.$$

The observed intensity of the green line is proportional to $A_2 n$ and

$$A_2 n = \frac{E A_2}{A_1 + A_2 + Z_Q}.$$

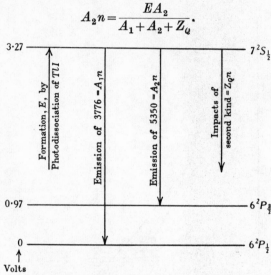

Fig. 57. Energy level diagram of Tl atom showing emission of 3776 and 5350 and impacts of the second kind.

Without foreign gas

$$A_2 n = \frac{E A_2}{A_1 + A_2},$$

whence the "quenching", Q, is

$$Q = \frac{A_1 + A_2}{A_1 + A_2 + Z_Q}$$

or

$$Q = \frac{1}{1 + \dfrac{Z_Q}{A_1 + A_2}} \qquad \text{......(146)}.$$

It is shown at the end of Chap. III ($\S 7b$) that

$$A_1 + A_2 = 7 \cdot 2 \times 10^7,$$

and therefore Z_Q and finally $\sigma_Q{}^2$ can be obtained from experimental measurements of Q.

The final results are shown in Table XXVIII. With I_2 and I as foreign gases, the dependence of σ_Q^2 on velocity is the same as that found by Terenin and Prileshajewa with excited sodium atoms; i.e. with I_2, σ_Q^2 decreases with increasing velocity, and with I there is substantially no change. The results, however, with TlI indicate just the reverse, namely an increase of σ_Q^2 with increasing velocity.

TABLE XXVIII

Optically dissociated molecule	Foreign gas	Velocity of excited Tl atom in cm./sec. $\times 10^{-5}$	$\sigma_Q^2 \times 10^{16}$
TlI	I_2	0·30	98
,,	,,	0·36	93
,,	,,	0·40	69
,,	I	0·35	45
,,	,,	0·36	37
,,	,,	0·40	27
,,	,,	0·46	40
,,	TlI	0·33	29
,,	,,	0·38	63
,,	,,	0·44	86

10. OTHER COLLISION PROCESSES

Collisions of the second kind involving sensitized fluorescence and chemical reactions have already been described in Chap. II, along with their interpretation in the light of quantum-mechanical principles. We are concerned in this chapter with only those experiments which yield a quantitative estimate of effective cross-sections associated with collisions of the second kind. Besides pressure-broadening collisions and quenching collisions there are three other types of collisions from experiments on which quantitative data may be obtained.

10a. COLLISIONS INVOLVING THE SODIUM TRANSITION $3\,^2P_{3/2} \rightarrow 3\,^2P_{1/2}$. It was first observed by Wood[109, 110] that, upon excitation of sodium vapour at low pressure by one of the D lines, only that line appeared as resonance radiation; whereas, upon introducing a foreign gas, or by raising the sodium vapour pressure, both D lines appeared. It is clear that the appearance of, say, the D_1 line, when sodium vapour mixed with argon is excited by D_2, must be due to a collision which lowers a sodium atom from the $3\,^2P_{3/2}$ state to the $3\,^2P_{1/2}$ state, that is, a collision of the type

$$\text{Na}\,(3\,^2P_{3/2}) + A = \text{Na}\,(3\,^2P_{1/2}) + A.$$

Similarly, by exciting with the D_1 line, in the presence of argon, the collision

$$\text{Na}\,(3\,^2P_{1/2}) + A = \text{Na}\,(3\,^2P_{3/2}) + A$$

takes place.

Lochte-Holtgreven[49] repeated Wood's experiment under better conditions, and measured the ratio of the intensities of the two D lines emitted by sodium vapour upon excitation first with D_2 and then with D_1, using four different foreign gases: argon, a mixture of neon and helium, nitrogen and hydrogen. Lochte-Holtgreven expected that, as the foreign gas pressure was increased, the ratio D_1/D_2 should approach $1/2$ upon excitation with D_2, and the ratio D_2/D_1 should approach 2 upon excitation with D_1. These expectations were confirmed within the limits of experimental error in the case of the inert gases. In the case of nitrogen and of hydrogen, the results were influenced by quenching. The resonance lamp was so constructed that a layer of unexcited sodium atoms lay between the emitting layer and the exit window, which, by absorbing the two D lines unequally, was partly responsible for the failure of the experimental ratios to reach completely their theoretical values. Control experiments showed the effect of this absorbing layer as well as the effect of diffused resonance radiation. In Fig. 58, the ratio of the intensity of the D_1 line to that of the D_2 line is plotted against the argon pressure, when the exciting light was D_2, and when the absorption within the resonance lamp was reduced to a minimum. It is seen that the ratio approaches the theoretical value $1/2$ at high argon pressures. From the initial portion of the curve it is possible to estimate the effective cross-section associated with the process $3\,^2P_{3/2} \rightarrow 3\,^2P_{1/2}$ caused by collisions with argon. The value of σ^2 for this process is very roughly 40×10^{-16}, whereas for the process $3\,^2P_{1/2} \rightarrow 3\,^2P_{3/2}$ it is roughly

18×10^{-16}. In the case of a neon-helium mixture, the two values of σ^2 are also approximately in the ratio of 2 to 1.

10*b*. COLLISIONS CONNECTED WITH PHOTO-IONIZATION. It was first shown by Mohler, Foote and Chenault[64] that an ionization current was established in caesium vapour when it was illuminated by various lines in the principal series of the caesium spectrum. Similar results were obtained by Lawrence

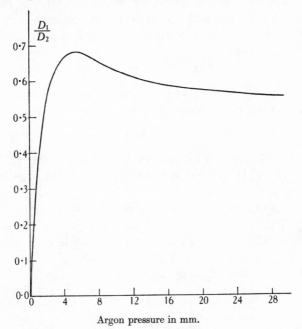

Argon pressure in mm.

Fig. 58. Effect of argon in causing the emission of D_1 when sodium vapour is illuminated with D_2.

and Edlefsen[45] with rubidium vapour illuminated by rubidium lines. It was at first assumed that an excited atom, formed by line absorption, received sufficient energy during a collision with a normal atom to ionize it. Thus:

$$Cs\,(m\,^2P) + Cs = Cs^+ + Cs + e,$$

where the $Cs\,(m\,^2P)$ atom is produced by the absorption of the caesium line $6\,^2S_{1/2}-m\,^2P$. Although this collision process is reasonable in the case of an atom excited to a state within a few hundredths of a volt of ionization (such as $15\,^2\mathrm{P}$ or higher), it is improbable when $m = 8$ or 9, since the number of collisions involving the requisite amount of energy is too small to account for the observed photo-ionization. The explanation therefore was advanced by Franck and Jordan that a collision between an excited atom and a normal atom resulted in the formation of a molecular ion and an electron. This explanation was borne out by further investigations of photo-ionization in caesium vapour by Mohler and Boeckner[66]. According to this point of view, the photo-ionization process consists of two parts, the collision process

$$\mathrm{Cs}\,(m\,^2\mathrm{P}) + \mathrm{Cs} = (\mathrm{Cs\,Cs})',$$

where the dash indicates a high electronic level of the molecule, and then the spontaneous process

$$(\mathrm{Cs\,Cs})' = (\mathrm{Cs\,Cs})^+ + e.$$

The first of these processes would be expected to depend on the pressure and the second to be independent of the pressure. On the basis of photo-ionization experiments in caesium vapour, Mohler and Boeckner were able to calculate roughly the product of the lifetime of a caesium atom, τ, and the effective cross-section, σ^2, associated with the collision in which an excited molecule is formed. The probability of ionization of an excited molecule, E_c, was found to be independent of pressure.

TABLE XXIX

m	$6\,^2\mathrm{S}_{1/2}-m\,^2\mathrm{P}$	E_c	$\tau\sigma^2 \times 10^{19}$
8	3888	0·003	0·22
9	3612	0·154	1·0
10	3477	0·26	1·1
11	3398	0·40	1·1
12	3347	0·50	1·0
13, 14	3300	0·77	1·2
16, 17	3250	0·89	3·3
20	3225	0·93	16
29	3200	1·00	48

The values of $\tau\sigma^2$ and E_c are given for nine different wavelengths in Table XXIX. Since the various τ's are not known, the absolute values of the effective cross-sections cannot be calculated. A rough estimate of τ, however, seems to indicate that σ is large compared with usual atomic dimensions.

In the presence of a foreign gas the photo-ionization was found to be diminished. This was explained by Mohler and Boeckner[67] on the basis of collisions of the type

$$\mathrm{Cs}\,(m\,^2\mathrm{P}) + F = \mathrm{Cs}\,(n\,^2\mathrm{P}) + F,$$

where F is a foreign gas molecule and $n\,^2\mathrm{P}$ is a lower excited state. From measurements of the diminution of photo-ionization by foreign gases, it was possible to estimate the effective cross-section associated with the above collision, and it was found in the case of argon, helium, nitrogen and hydrogen to be very nearly the gas-kinetic value.

10c. COLLISIONS INVOLVING THE ENHANCEMENT OF SPARK LINES. The statement was made in Chap. I, in connection with sources for exciting resonance radiation, that an inert gas, in which an electric discharge is maintained, is very effective in exciting the atoms of an admixed vapour. This phenomenon was described at greater length in Chap. II under the heading "Sensitized Fluorescence". The collision process in question is an example of a collision of the second kind between an excited inert gas atom and a normal atom, in which the normal atom is raised to a level from which it can radiate. Upon comparing the spectral lines emitted by the admixed vapour with the normal arc spectrum of the vapour, a qualitative estimate of the effectiveness of the sensitizing collisions is obtained. Recently, a series of experiments have been performed by Duffendack and Thomson[16] which can be interpreted quantitatively and which yield relative values of effective cross-sections. In these experiments, the relative intensities of certain spark lines of silver, gold, aluminium and copper, emitted by reason of impacts with helium and neon ions, was measured and compared with the relative intensities of the same lines emitted by a condensed spark. The ratio of the intensities of a group of lines originating from the same upper level under these two conditions was called the "enhancement", and was used as a measure of the effectiveness of the sensitizing collision. A typical collision of this type is as follows:

$$\mathrm{Ne^+} + \mathrm{Cu} = \mathrm{Ne} + \mathrm{Cu^+}\,(^3\mathrm{D}_1),$$

and Table XXX explains how the enhancement for this collision was calculated.

TABLE XXX

Copper line	Initial level	Normal intensity	Intensity in neon	Enhancement
2485·95	$(d^9s)^3D_1$	4·3	57	13·2
2590·68	,,	2·1	32	15·2
2703·34	,,	3·8	57	15·0
2721·84	,,	1·9	27	14·7
				(Aver.) 14·6

In Tables XXXI and XXXII are given a number of results which will be discussed from the standpoint of quantum theory in the next section.

TABLE XXXI

COPPER ION LEVELS EXCITED BY NEON IONS

Copper ion level $Cu^+(d^9s)$	Relative enhancement by neon ions	Energy discrepancy in volts
$^3D_3$	1·0	0·40
$^3D_2$	2·7	0·36
$^3D_1$	14·6	0·16
$^1D_2$	9·7	0·12

TABLE XXXII

ALUMINIUM ION LEVELS EXCITED BY NEON IONS

Aluminium ion level	Relative enhancement by neon ions	Energy discrepancy in volts
5^1P	12·2	0·022
5^3P	>30	0·003
4^1F	1·0	0·27
4^3F	2·6	0·28

11. THEORETICAL INTERPRETATION OF QUENCHING COLLISIONS

11a. GENERAL PRINCIPLES. Collision processes involving an interchange of excitation energy between the two colliding particles have been of considerable interest to theoretical physicists, since the publication in 1929 of an important paper on the subject by Kallmann and London [32]. Later papers by Morse and Stueckelberg [69], Rice [82, 83], Zener [117], London [50], Landau [43, 44], Morse [70], and Stueckelberg [94], have all contained applications of various methods of the quantum mechanics to collisions of this type, with varying degrees of success, depending upon the validity of the assumptions that were made. A brief survey of the various collision processes that have been dealt with in the preceding sections and in Chap. II will show such a wide range in regard to type, nature of interaction between colliding partners, amount of energy transferred, etc., that it is not surprising that the whole range has not been embraced completely by one theory. Certain general principles, however, stand out as being fundamental in all these treatments, and will be found to be adequate to interpret some of the existing experimental results. From a quantum-mechanical standpoint, a collision of the second kind is a special case of the general problem of the molecule. A collision is regarded as the temporary formation of a quasi-molecule, and the Franck-Condon curves of this molecule, representing conditions before and after collision, play the main rôle in the theoretical calculation. In general, the effective cross-section associated with a particular collision process depends upon (1) the relative kinetic energy before impact, (2) the law of interaction between the two systems, and (3) the difference between the relative kinetic energies before and after impact.

The dependence of effective cross-section upon relative kinetic energy before impact is shown in Figs. 55 and 56, and in Table XXVIII, in connection with excited sodium and excited thallium atoms. It is seen that the effective cross-section at first decreases, and then seems to increase as the velocity of the excited atom increases. These curves are the analogue of the Ramsauer curves for electrons. According to Morse [70], the decreasing part of these curves is in agreement with theory, but the theory does not predict a rise in effective cross-section at high velocities. Instead, it is consistent with the theory that, at low velocities, σ^2 should attain a maximum, and at still lower velocities, decrease. This effect has not yet been observed. Both the theory and the experimental results are in too undeveloped a state to warrant any further discussion.

It has already been mentioned in Chap. II that the conditions favouring a collision of the second kind are (1) the difference in the relative kinetic energies before and after impact (which must equal the difference between the energy that one system has to give up and the energy that the other system can take, and which is therefore termed by Duffendack the "energy discrepancy") should be as small as possible, and (2) the total spin angular momentum of both systems should be conserved (known as Wigner's rule). The experiments of Cario, Beutler, and many others, discussed in Chap. II, illustrate how collisions giving rise to sensitized fluorescence satisfy these principles. The method of sensitized fluorescence is particularly well adapted to the study of these points, because it allows only the energy discrepancy to be varied, while the other factors remain fairly constant. The energy discrepancy can be accurately calculated in such experiments, because the colliding systems are atoms whose energy levels are completely known. In collisions involving sensitized fluorescence, however, it is difficult to estimate effective cross-sections. The results given in Chap. II, therefore, had to be discussed in a qualitative manner.

11b. ENHANCEMENT OF COPPER AND ALUMINIUM IONIC LEVELS. The experiments of Duffendack and Thomson on the enhancement of copper and aluminium spark lines by neon ions, described in §10c, allow a more quantitative interpretation than the experiments on sensitized fluorescence in Chap. II. The "enhancement", as defined by Duffendack and Thomson, can be regarded as roughly proportional to the effective cross-section associated with the collision process

$$Ne^+ + Cu = (Cu^+)' + Ne.$$

In the third column of Tables XXXI and XXXII are given the values of the energy discrepancy corresponding to each ionic level that was excited. The relation between enhancement and energy discrepancy is shown graphically in Fig. 59.

It is clear from the figure that the triplet levels of both copper and aluminium are excited more strongly than the corresponding singlet levels. Taking all the triplet levels together, it is seen that the enhancement increases as the energy discrepancy decreases. The same is true for all the singlet levels. These two results are in accord with all previous work on sensitized fluorescence. The preference of the triplet levels over singlet levels with smaller energy discrepancy is, however, not in

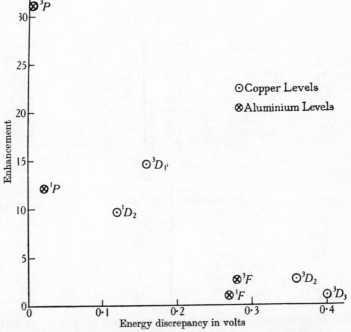

Fig. 59. Relation between enhancement and energy discrepancy.

accord with Wigner's rule concerning conservation of spin angular momentum. It is difficult to say whether this constitutes a serious objection to Wigner's rule or not.

11c. ENERGY INTERCHANGE WITH MOLECULES. It has been mentioned that, in experiments on sensitized fluorescence, although the energy discrepancy can be determined unambiguously, the absolute value of the effective cross-section cannot. Just the opposite is true of experiments on the quenching of resonance radiation. Although the effective cross-sections associated with the quenching of mercury, sodium and thallium resonance radiation can be regarded as known with a fair degree of accuracy, it is not always possible to decide precisely what transition the excited atom or the foreign gas molecule performs. One reason for this lies in our ignorance of the energy levels of some molecules. The vibrational levels of the higher electronic states of the NaI, TlI and

TABLE XXXIII

Atom	Transition	Energy in volts	Transition	Energy in volts
Hg	$6\,^3P_1 \to 6\,^3P_0$	0·218	$6\,^3P_1 \to 6\,^1S_0$	4·862
Na	—	—	$3\,^2P \to 3\,^2S_{1/2}$	2·094
Tl	$7\,^2S_{1/2} \to 6\,^2P_{3/2}$	2·306	$7\,^2S_{1/2} \to 6\,^2P_{1/2}$	3·267
Cd	$5\,^3P_1 \to 5\,^3P_0$	0·0669	$5\,^3P_1 \to 5\,^1S_0$	3·783
I	$^2P_{1/2} \to {}^2P_{3/2}$	0·758	$^4P_{5/2} \to {}^2P_{3/2}$	6·741
Br	$^2P_{1/2} \to {}^2P_{3/2}$	0·455	$^4P_{5/2} \to {}^2P_{3/2}$	7·82

TABLE XXXIV

Molecule	Energy of dissociation into normal atoms in volts	Molecule	Energy of dissociation into normal atoms in volts
H_2	4·44	NaH	2·25
O_2	5·09	CdH	0·67
H_2O	5·05	HgH	0·369
	$(H_2O = OH + H)$	NaI	2·98
I_2	1·54	TlI	2·61
Br_2	1·96	NaBr	3·66

NaBr molecules, for example, are unknown. There are some cases where two or three different processes seem to be equally suited to explain quenching. In these cases, there is always the possibility that all the processes occur, and that the measured effective cross-section is only an average value. In order to calculate the energy discrepancy in quenching collisions by molecules, it is necessary to know the energy that the excited atom can give up, and, in some cases, the energy necessary to dissociate the molecule. These quantities are given for convenient reference in Tables XXXIII and XXXIV.

It is also necessary to know the energies of those vibrational levels of a molecule which lie nearest to some given value. These will be found throughout the next four tables. Knowing the heat of dissociation and the energies of various vibrational levels of the hydrogen molecule, Kaplan[33] was able to explain the experiments of Bonhoeffer[9] and Mohler[65] on the excitation of spectral lines by recombining hydrogen atoms. These experiments indicate that, when two hydrogen atoms combine to form H_2 in the presence of various metals, certain spectral lines of these metals are excited. In most cases it was possible to connect the intensity of a spectral line with the difference between the heat of recombination and the energy of a vibrational level. This difference was the energy effective in exciting the line. This energy minus the excitation energy of the line then constituted the energy discrepancy. In agreement with the ideas of quantum mechanics, the smaller the energy discrepancy was, the larger was found the intensity of the line.

11d. COLLISIONS WITH EXCITED MERCURY ATOMS. The interpretation of experiments on the quenching of mercury resonance radiation may be best discussed in connection with Table XXXV, in which the various possible quenching processes are listed. The behaviour of CH_4, CO and N_2 seems to indicate that the mercury atom is lowered to the $6\,^3P_0$ state in causing the transition of the molecule from the zero vibrational state to the first vibrational state of the normal electronic level. None of the molecules has an electronic level near 4·86 volts, and, although a high vibrational level may lie near this value, a transition to such a high level does not seem probable. In the case of quenching by hydrogen, on the other hand, it appears that the mercury atom gives up all of its energy. A complete discussion of this point was given in Chap. II.

In the case of H_2O, three possibilities are shown in the table. The first certainly exists, because H_2O molecules are known to be very effective in producing metastable mercury atoms. The other two processes are energetically possible. The behaviour of NO is very interesting, in that its effective cross-section is the second largest listed in the table. This value was obtained by Bates, who considers that both processes listed in the table are possible. This view was confirmed by the work of Noyes[73].

As regards the quenching of CO_2 and NH_3, there is not sufficient evidence to enable one to distinguish between the processes listed in the table. The explanation of the quenching of mercury resonance radiation by O_2 is still in doubt. Of the

TABLE XXXV

[Hg' denotes 6^3P_1; Hg_m denotes 6^3P_0; ( )$_v$ denotes a vibrational level of the normal electronic state; chemical symbols alone denote the normal state.]

Foreign gas molecule	Possible quenching process	Energy available volts	Energy required volts	Energy discrepancy volts	σ_q^2 cm.^{2} $\times 10^{16}$
CH_4	$Hg' + CH_4 = Hg_m + (CH_4)_{v1}$	·218	·161	$+ ·057$	·0852
CO	$Hg' + CO = Hg_m + (CO)_{v1}$	·218	·265	$- ·047$	5·82
N_2	$Hg' + N_2 = Hg_m + (N_2)_{v1}$	·218	·288	$- ·070$	·274
H_2	$Hg' + H_2 = Hg + H + H$	4·86	4·44	$+ ·42$	8·60
H_2O	$Hg' + H_2O = Hg_m + (H_2O)_{v1}$	·218	·197	$+ ·021$	1·43
,,	$Hg' + H_2O = Hg + OH + H$	4·86	5·05	$- ·19$	1·43
,,	$Hg' + H_2O = HgH + OH$	4·86 $+ 0·37$	5·05	$+ ·18$	1·43
NO	$Hg' + NO = Hg_m + (NO)_{v1}$	·218	·231	$- ·013$	35·3
,,	$Hg' + NO = Hg + (NO)_v$	4·86	4·90	$- ·04$	35·3
CO_2	$Hg' + CO_2 = Hg_m + (CO_2)_{v1}$	·218	·238	$- ·020$	3·54
,,	$Hg' + CO_2 = Hg + (CO_2)_v$	4·86	5·50	$- ·64$	3·54
NH_3	$Hg' + NH_3 = Hg_m + (NH_3)_{v1}$	·218	·202	$+ ·016$	4·20
,,	$Hg' + NH_3 = Hg + (NH_3)'$	4·86	4·90(?)	$- ·04$	4·20
O_2	$Hg' + O_2 = Hg + O + O$	4·86	5·09	$- ·23$	19·9
,,	$Hg' + O_2 = Hg + O_2\,(^3\Sigma)$	4·86	4·86	0	19·9
,,	$Hg' + O_2 = Hg + O_2\,(^1\Sigma)$	4·86	4·90	$- ·04$	19·9
,,	$Hg' + O_2 = (Hg'O_2)$	—	—	—	19·9
C_6H_6	$Hg' + C_6H_6 = Hg + C_6H_5 + H$	4·86	—	—	60

four processes listed in the table the second was suggested by Mitchell[63], the third by Bates[7] and the fourth by Noyes[72]. Some of these processes were discussed in Chap. II in connection with the formation of ozone. The quenching ability of C_6H_6 was explained by Bates as being due to the removal of a hydrogen atom.

In the case of those molecules which are either known to produce metastable mercury atoms, or which have a first vibrational state near enough to 0·218 volt to make this a possibility, the variation of effective cross-section with energy discrepancy is instructive. In Fig. 60, the effective cross-section is plotted against the energy of the first vibrational state. No curve can be drawn through the points because the

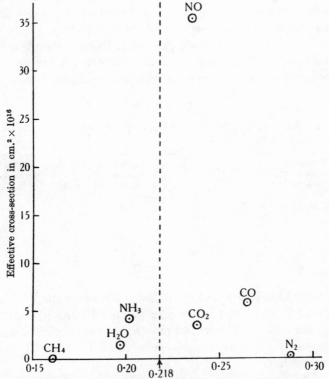

Fig. 60. Relation between effective cross-section and energy discrepancy.

law of interaction between the excited mercury atom and the foreign molecule is different in each case. The effective cross-section, therefore, is not a function of the energy discrepancy *only*. Nevertheless the points in Fig. 60 show quite definitely that large effective cross-sections are associated with small energy discrepancies.

11*e*. COLLISIONS WITH EXCITED CADMIUM ATOMS. In regard to the quenching of excited cadmium atoms by N_2 and

CO, one can be fairly certain of the collision processes that take place. These are shown in Table XXXVI.

It is not reasonable to regard the cadmium transition $5\,^3P_1 \to 5\,^3P_0$, which must, of course, take place frequently, as a quenching process, in view of the fact that the converse transition $5\,^3P_0 \to 5\,^3P_1$ also takes place very often. In order that cadmium resonance radiation be quenched, it is quite likely that the $5\,^3P_1$ atoms must be brought all the way to the normal level. Two possibilities are present to explain the quenching ability of H_2. The first, involving the formation of the CdH molecule, is almost certain to take place, not only because the

TABLE XXXVI

[Cd′ denotes $5\,^3P_1$, Cd denotes $5\,^1S_0$, ( )$_v$ denotes a vibrational level of the normal electronic state.]

Foreign gas molecule	Possible quenching process	Energy available volts	Energy required volts	Energy discrepancy volts
N_2	$Cd' + N_2 = Cd + (N_2)_v$	3·783	—	—
CO	$Cd' + CO = Cd + (CO)_v$	3·783	—	—
H_2	$Cd' + H_2 = CdH + H$	3·783 + ·67	4·44	+ ·013
H_2	$Cd' + H_2 = Cd + (H_2)_v$	3·783	About 3·84	− ·06

energy discrepancy is so small, but also because the CdH band spectrum was observed by Bender in mixtures of cadmium vapour and H_2. The second process has been included merely because it is energetically possible, with a small energy discrepancy.

11f. COLLISIONS WITH EXCITED SODIUM ATOMS. The possible quenching processes involved in collisions between excited sodium atoms and foreign gas molecules are shown in Table XXXVII. The behaviour of N_2, CO and CO_2 is fairly certain. N_2 and CO have vibrational levels in the neighbourhood of 2·094 volts, and it is reasonable to assume the same for CO_2. According to the energy discrepancies, CO ought to have a larger effective cross-section than N_2, but this is not the case. The disagreement, however, is not serious. With H_2, the two processes listed in the table are about equally reasonable from the standpoint both of possibility and of energy discrepancy. Of the three processes listed for I_2, the first is the least reasonable. The vibrational levels of the first excited electronic level of I_2 lie between 1·92 and 2·77 volts, so that the energy dis-

TABLE XXXVII

[Na′ denotes $3\,^2P$; $(\ )_v$ and $(\ )_v'$ denote a vibrational level of the normal and of an excited electronic state respectively; a chemical symbol alone denotes the normal state. I_m refers to the metastable $^2P_{1/2}$ state.]

Foreign gas molecule	Possible quenching process	Energy available volts	Energy required volts	Energy discrepancy volts	σ_0^2 cm.$^2 \times 10^{16}$
N_2	$Na' + N_2 = Na + (N_2)_v$	2·094	1·94 or 2·20	+ ·15 or – ·11	~6
CO	$Na' + CO = Na + (CO)_v$	2·094	2·03 or 2·26	+ ·06 or – ·17	~4
CO_2	$Na' + CO_2 = Na + (CO_2)_v$	2·094	—	—	~15
H_2	$Na' + H_2 = Na + (H_2)_v$	2·094	1·93 or 2·34	+ ·16 or – ·25	~6
,,	$Na' + H_2 = NaH + H$	2·094 + 2·25	4·44	– ·10	~6
I_2	$Na' + I_2 = Na + I_m + I$	2·094	1·54 + 0·76	– ·21	~40
,,	$Na' + I_2 = Na + (I_2)_v'$	2·094	Between 1·92 and 2·77	—	~40
,,	$Na' + I_2 = (NaI)_v' + I$	2·094 + 2·98	1·54 + ?	—	~40
Br_2	$Na' + Br_2 = Na + Br + Br$	2·094	1·96	+ ·13	~100
,,	$Na' + Br_2 = Na + (Br_2)_v'$	2·094	Between 1·93 and 2·39	—	~100
,,	$Na' + Br_2 = (NaBr)_v'$	2·094 + 3·66	1·96 + ?	—	~100
I	$Na' + I = (NaI)_v'$	2·094 + 2·98	?	—	~40

crepancy in the second process would presumably be quite small. The third process is preferred by Terenin and Prileshajewa, but without a knowledge of the vibrational levels of the excited electronic states of NaI the energy discrepancy cannot be calculated. Similarly, in the case of Br_2, the second and third processes are more reasonable than the first, for reasons similar to those given for I_2. In the case of I, nothing of a definite character can be said, except that a theoretical calculation of the effective cross-section to be expected for such a process, made by Terenin and Prileshajewa[97], yields a value about a million times smaller than the measured value.

11g. COLLISIONS WITH EXCITED THALLIUM ATOMS. The possible collisions between excited thallium atoms and I_2 and I are given in Table XXXVIII.

Of the four processes listed in connection with I_2, the first is preferred, because of its extraordinarily small energy discrepancy. The second process is also very reasonable and would

TABLE XXXVIII

[Tl′ denotes $7\,^2S_{1/2}$; Tl_m denotes $6\,^2P_{3/2}$; I_m denotes the metastable $^2P_{1/2}$ state; ( )$_v'$ refers to a vibrational level of an excited electronic state.]

Foreign gas molecule	Possible quenching process	Energy available volts	Energy required volts	Energy discrepancy volts	σ_q^2 cm.$^2 \times 10^{16}$
I_2	$Tl' + I_2 = Tl_m + I_m + I$	2·306	1·54 +·758	+·008	~70
,,	$Tl' + I_2 = Tl + I_m + I_m$	3·267	1·54 +·76+·76	+·21	~70
,,	$Tl' + I_2 = Tl_m + (I_2)_v'$	2·306	Between 1·92 and 2·77	—	~70
,,	$Tl' + I_2 = (TlI)_v' + I$	3·267 +2·61	1·54 +?	—	~70
I	$Tl' + I = (TlI)_v'$	3·267 +2·61	?	—	~30

presumably have a very small energy discrepancy. The third and fourth processes involve the emission of a band spectrum as fluorescence. The fact that Prileshajewa did not observe any fluorescent bands is an objection to these processes. This point, however, is not completely settled. In connection with the behaviour of I, the same objection that was made before for this kind of process holds here. Theoretically, it should have an effective cross-section of about 10^{-22} cm., whereas the measured value is about a million times larger.

12. RAPIDITY OF ESCAPE OF DIFFUSED RESONANCE RADIATION FROM A GAS

12a. EXPERIMENTS WITH MERCURY VAPOUR AT LOW PRESSURES. In Chap. III the experiments of Webb and

Messenger[101] and those of Garrett[23] were described in connection with their significance as measurements of the lifetime of the $6\,{}^3P_1$ state of the mercury atom. It was emphasized that the decay constant of the exponential curve, representing the decay of the resonance radiation emitted by mercury vapour, could be interpreted as the Einstein A coefficient $(1/\tau)$, only if the vapour pressure was so low that the diffusion of the resonance radiation, through repeated absorptions and emissions, could be neglected. When this is not the case, the lifetime of

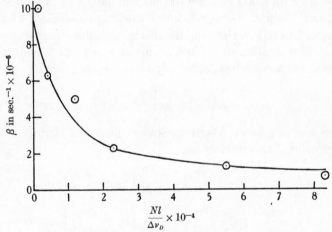

Fig. 61. Relation between decay constant of escaping mercury resonance radiation and vapour pressure.

(Vapour pressure range from 0 to about 0·001 mm.)

the radiation escaping from the mercury vapour is considerably longer than the lifetime of an atom. This is shown in Fig. 61, where the exponential constant of decay of the radiation, β, in sec.$^{-1}$, is plotted against the expression $Nl/\Delta\nu_D$, where N is the number of absorbing atoms, l the thickness of the layer of mercury vapour and $\Delta\nu_D$ the Doppler breadth of the diffusing radiation. Since l is constant and $\Delta\nu_D$ varies only slightly in the temperature range covered by Webb and Messenger's results, the quantity $Nl/\Delta\nu_D$ is very nearly proportional to the vapour pressure. The decrease in β as the vapour pressure increases from zero to about 0·001 mm. indicates the increase in the lifetime of the escaping radiation. To explain these results it is necessary to make use of Milne's theory, which was introduced in § 8 of this chapter.

12b. MILNE'S THEORY. Imagine an infinite slab of gas bounded by the planes $x = 0$ and $x = l$. Suppose the gas has been excited for a while and, when $t = 0$, there is a distribution of excited atoms, n_2, depending on x, thus $n_2 = f(x)$. Due to repeated absorptions and emissions from moving atoms, the radiation diffusing through the gas will have a Doppler distribution. Calling the equivalent absorption coefficient of the gas for this radiation $\bar{k}$, and the lifetime of the excited atoms τ, the concentration of excited atoms at any point and at any time after the excitation has been removed is given by

$$\frac{\partial^2}{\partial x^2}\left(n_2 + \tau\frac{\partial n_2}{\partial t}\right) = 4\bar{k}^2\tau\frac{\partial n_2}{\partial t} \qquad \ldots\ldots(147),$$

and the net forward and backward fluxes of radiation are represented respectively by

$$\pi I_+ = \frac{\pi\sigma'}{n_1}\left[\left(n_2 + \tau\frac{\partial n_2}{\partial t}\right) - \frac{1}{2\bar{k}}\frac{\partial}{\partial x}\left(n_2 + \tau\frac{\partial n_2}{\partial t}\right)\right]\ldots\ldots(148)$$

and $$\pi I_- = \frac{\pi\sigma'}{n_1}\left[\left(n_2 + \tau\frac{\partial n_2}{\partial t}\right) + \frac{1}{2\bar{k}}\frac{\partial}{\partial x}\left(n_2 + \tau\frac{\partial n_2}{\partial t}\right)\right]\ldots\ldots(149).$$

In these expressions $\partial n_2/\partial t$ represents the total rate of change of excited atoms due to the three Einstein processes, since there is no other process of formation *after* the cut-off of the excitation, and no other process of decay in the absence of impacts of the second kind. The problem is to calculate πI_+ at $x = l$ on the basis of the boundary conditions

    (1) when $t = 0$, $n_2 = f(x)$;

    (2) when $t > 0$, $I_+ = 0$ at $x = 0$;

    (3) when $t > 0$, $I_- = 0$ at $x = l$.

It was shown by Milne[59] that the radiation escaping from the face $x = l$ could be represented by a series of the form

$$\pi I_+(x = l) = A_1 e^{-\beta_1 t} + A_2 e^{-\beta_2 t} + \ldots \quad \ldots\ldots(150),$$

where the A's depend upon the original distribution of excited atoms when $t=0$, and the β's are given by

$$\beta_i = \frac{1/\tau}{1+\left(\dfrac{\bar{k}l}{\lambda_i}\right)^2} \qquad \ldots\ldots(151),$$

where λ_i is the ith root of the equation

$$\tan y = \frac{\bar{k}l}{y} \qquad \ldots\ldots(152).$$

It can be shown that β_2, β_3, etc., are all larger than β_1, and that $\beta_2 t$, $\beta_3 t$, etc., are so large when t is of the order of 10^{-4} sec. or more, that all terms except the first can be neglected. Milne's

TABLE XXXIX

Temp. °K.	$\Delta\nu_D \times 10^{-9}$ $=5\cdot97\times10^7\sqrt{T}$	$N \times 10^{-13}$	$\dfrac{Nl}{\Delta\nu_D}\times10^{-4}$	$k_0 l = 1\cdot33\,\dfrac{Nl}{\Delta\nu_D}$ $\times10^{-4}$	$\bar{k}l$ (Samson)	$\beta\times10^{-6}$ Theor.	$\beta\times10^{-6}$ Exp.
254	·95	·078	·15	·20	·16	8·0	10
263	·97	·22	·41	·55	·38	6·5	6·3
273	·99	·66	1·2	1·6	1·0	3·8	5
280	1·00	1·3	2·3	3·1	1·8	2·3	2·3
290	1·02	3·1	5·5	7·3	2·9	1·3	1·3
295	1·03	4·7	8·3	11·1	3·5	1·05	·75

result is therefore as follows: After a time has elapsed, the radiation escaping from a gas decays exponentially with the time, with an exponential constant, β, given by

$$\beta = \frac{1/\tau}{1+\left(\dfrac{\bar{k}l}{\lambda_1}\right)^2} \qquad \ldots\ldots(153),$$

where λ_1 is the *first* root of Eq. (152).

In order to compare Webb and Messenger's results with Milne's theory, it is necessary to calculate $\bar{k}l$ at the various vapour pressures in the experimental range. This is done by Samson's method and is described in detail in §8c. Table XXXIX contains the theoretical values of β along with Webb and Messenger's experimental results, and the heavy curve in Fig. 62 is a graph of the theoretical results. It must be emphasized that the experimental tube of Webb and Messenger was only a very rough experimental approximation to an infinite slab, and, in the absence of an accurate knowledge of the thickness of the mercury vapour layer, a value of l equal to 1·8 cm. was chosen for purposes of calculation. The agreement is satisfactory in view of the lack of correspondence

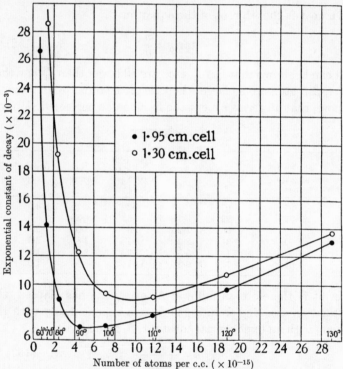

Fig. 62. Decay of Hg resonance radiation escaping from mercury vapour.
(Vapour pressure range from 0·01 to 1 mm.)

between the experimental conditions and those postulated by the theory. The discrepancy between the experimental and theoretical values of β at the highest temperature, where $k_0 l$ is 11·1, may be due partly to the inadequacy of Samson's method of calculating the equivalent opacity at this vapour pressure.

12c. EXPERIMENTS WITH MERCURY VAPOUR AT HIGHER PRESSURES. The rapidity of escape of mercury resonance radiation from a slab of mercury vapour after the cut-off of the excitation, was measured by Webb [100] and by Hayner [27] by an electrical method. A repetition of this experiment with optical excitation was made by Zemansky [111] for vapour pressures ranging from about 0·01 mm. to 1 mm., and with two different thicknesses of the mercury layer, 1·95 cm. and 1·30 cm. The results of Zemansky indicated a rapid increase in the lifetime of the radiation in the pressure range 0·01 to about 0·3 mm., and then a slower decrease in lifetime from 0·3 to 1 mm. In terms of β, the exponential constant of decay, this means first a decrease and then an increase, as shown in Fig. 62. The interpretation of these experiments is not yet certain. A discussion of the left-hand part of the curves (decreasing β), separate from that of the right-hand part, will tend to point out more clearly the nature of the problem. In regard to this left-hand part, there are two theories: (1) that the fundamental process at hand is the diffusion of resonance radiation composed of those frequencies that are absorbed and emitted by virtue of the Doppler effect; (2) that collisions between excited $6\,^3P_1$ atoms and normal atoms produce metastable $6\,^3P_0$ atoms which, in diffusing through the mercury vapour, either reach the walls and give up their energy, or are knocked up again to the $6\,^3P_1$ state from which they radiate.

On the basis of the first theory, Milne's theory gives the same result as in Webb and Messenger's experiments, namely

$$\beta = \frac{1/\tau}{1 + \left(\dfrac{\bar{k}l}{\lambda_1}\right)^2} \qquad \ldots\ldots(154),$$

with the distinction that $\bar{k}$, in this case, is the equivalent absorption coefficient at *high* vapour pressure, where Samson's method of calculation would not be expected to hold, and λ_1, under these conditions, is very nearly equal to $\pi/2$.

12*d*. EQUIVALENT OPACITY AT HIGH PRESSURE. At high pressures, k can be calculated by a method due to Kenty [34, 116] which, in contradistinction to Samson's method, breaks down at low pressures. According to Kenty, the motions of both emitting and absorbing atoms are assumed to have a Maxwellian distribution. The radiation diffusing through a gas is then found to have a diffusion coefficient, D, given by

$$D = \frac{\sqrt{2}\,l}{3\sqrt{\pi}\tau k_0}\left[\sqrt{2}\,F\left(\sqrt{\ln k_0 l}\right) - F\left(\sqrt{2\ln k_0 l}\right)\right]\quad\ldots\ldots(155),$$

where τ is the lifetime of the excited state, l the thickness of the layer of gas, k_0 the absorption coefficient at the centre of the line, and

$$F(t) = e^{-t^2}\int_0^t e^{x^2}\,dx \qquad \ldots\ldots(156).$$

TABLE XL

Temp. ° K.	$N \times 10^{-15}$	$\dfrac{Nl}{\Delta\nu_D} \times 10^{-6}$	$k_0 l$	$\bar{k}l$ (Kenty)	β Theor.	β Exp.
333	·770	1·38	183	30	29500	26600
343	1·40	1·64	218	33	24500	28100
343	1·40	2·46	327	42	15000	14200
353	2·50	2·90	386	47	12000	19300
353	2·50	4·35	578	58	7940	8810
363	4·40	5·02	668	63	6750	12100
363	4·40	7·53	1000	78	4380	7070

On the basis of Milne's theory, the diffusion coefficient of radiation for which the absorption coefficient is $\bar{k}$ is equal to

$$\frac{1}{4\bar{k}^2\tau} \qquad \ldots\ldots(157).$$

Equating the two expressions for the diffusion coefficient, the equivalent opacity $\bar{k}l$, at high pressures, is found to be

$$\bar{k}l = \sqrt{\frac{3}{4}}\sqrt{\frac{\pi}{2}}\cdot\sqrt{\frac{k_0 l}{\sqrt{2}\,F\left(\sqrt{\ln k_0 l}\right) - F\left(\sqrt{2\ln k_0 l}\right)}} \quad\ldots\ldots(158).$$

A table of values of $\bar{k}l$ for many values of $k_0 l$ will be found in the Appendix. In Table XL, the experimental values of β (corresponding to the left-hand part of the experimental curves) are compared to the theoretical values calculated by Milne's theory with the aid of Kenty's equivalent opacity. The agreement is seen to be satisfactory enough at least to lend credence to the theory of radiation diffusion.

Now, on the basis of the theory of metastable atom diffusion, it was shown by Zemansky [113] that the left-hand part of the curve could be represented by

$$\beta = \frac{\text{const.}}{\sigma_d{}^2 N} \qquad\qquad \ldots\ldots(159),$$

where the constant depends on the velocity of the atoms and the geometry of the layer of vapour, and $\sigma_d{}^2$ can be regarded as an effective cross-section for diffusion. This was shown to agree well with the shape of the curve, and enabled a value of σ_d to be inferred from the experiments. The resulting value, $2\cdot3 \times 10^{-8}$ cm., is remarkable in that it is smaller than the usual "gas-kinetic value", $3\cdot6 \times 10^{-8}$ cm. This result, by itself, does not constitute a serious objection to the metastable atom theory, since effective cross-sections for various processes depend upon the nature of the processes and can be expected to vary from very small to very large values. A much more serious objection to the metastable atom theory was pointed out by Kenty [34], namely, the doubt as to the existence of sufficient metastable atoms to give rise, by being raised to the $6\,^3P_1$ state, to a measurable amount of radiation. The *formation* of a metastable atom would depend upon a collision of the type

$$\text{Hg}\,(^3P_1) + \text{Hg} = \text{Hg}\,(^3P_0) + \text{Hg},$$

and the production of radiation would involve the two processes

$$\text{Hg}\,(^3P_0) + \text{Hg} = \text{Hg}\,(^3P_1) + \text{Hg},$$
$$\text{Hg}\,(^3P_1) = \text{Hg} + h\nu.$$

For the transition $^3P_1 \rightarrow {}^3P_0$, $0\cdot218$ volt of excitation energy must go into relative kinetic energy of the colliding partners, since a normal mercury atom has no energy level lower than $4\cdot7$ volts. In other words, the ability of normal mercury atoms to effect this transition ought to be about the same as the ability on the part of inert gas atoms to perform the same transition. From experiments on the quenching of mercury resonance radiation, it is found that the inert gases do not quench at all, or if they do, to a very small extent. Moreover, from theoretical considerations, the probability of any collision involving the transfer of $0\cdot218$ volt of excitation energy into the kinetic energy of the gas is expected to be extremely low. Substantially the same objections may be made in connection with the transition $^3P_0 \rightarrow {}^3P_1$, so that it appears extremely doubtful whether the experiments can be explained on the basis of the metastable atom theory. A final decision, however, cannot be made without further experimental work.

In reference to the right-hand portion of the experimental curves for β, there are also two possible theories: (1) impacts destroying either $^3P_1$ atoms or $^3P_0$, and (2) Holtsmark broadening of the absorption line giving rise to a smaller value of $\bar{k}l$ than is given by Kenty's formulas, thereby causing β to increase. It is impossible to decide which of these points of view is to be preferred, since they both seem equally suited to account for the somewhat slow rise in β that is observed. The whole matter must be left open until further experiments are performed, and further theoretical calculations on the effect of Holtsmark broadening upon the equivalent opacity are made.

The difficulties that arise in the interpretation of experiments on the decay of radiation from a gas appear also when an attempt is made to explain the large currents found at considerable distances from the end of a noble gas discharge. According to Kenty[35], resonance radiation is capable of diffusing through the gas at a much faster rate than was formerly supposed, because the equivalent opacity of the gas for the resonance radiation, as calculated on the basis of his theory, was small. The observed currents therefore could be interpreted as photoelectric currents. An equivalent interpretation, however, has been given by Found and Langmuir[20] on the basis of metastable atoms. Since this subject is beyond the scope of this book, the reader is referred to the original papers for a more complete discussion.

13. DIFFUSION AND COLLISIONS OF METASTABLE ATOMS

13*a*. EARLY WORK. The first measurements of the lifetime of metastable atoms were made by Meissner[55] and by Dorgelo[14]. An inert gas in an absorption tube was electrically excited for a while, and then the excitation was stopped. After a short time had elapsed, a beam of light, capable of being absorbed by metastable atoms (thereby raising the metastable atoms to a higher energy level), was sent through the absorption tube. The length of time beyond which absorption was no longer perceptible was measured, and called roughly the lifetime of the metastable atoms. Various methods of starting and stopping the excitation of the absorbing gas, and of starting and stopping the emitting lamp, were employed. They all had the disadvantage that the time between the cut-off of the excitation and the passage of the light could not be determined accurately enough to yield a reliable curve of decay of absorption against time. An improvement was made by Dorgelo and Washington [15], in that both the absorption tube and the emission lamp were operated on A.C. with a constant phase difference of 180°. The time between the cut-off of the excitation and the absorption of the light was varied by altering the frequency of the A.C. These qualitative experiments indicated that metastable inert gas atoms lasted approximately a few thousandths of a second after the excitation was removed, and that the lifetime depended on the gas pressure and temperature.

Since these early experiments, there have been many investigations on metastable atoms, which have not only been of interest in themselves, but have yielded information of the utmost importance in explaining phenomena occurring in gas discharges. Before these later experiments can be explained, it is advisable to consider first the general theory of the method.

13b. THEORY OF MEASUREMENT WITH INERT GASES. After considering and trying out various methods of exciting an inert gas, and of allowing a beam of light to traverse this gas after a known time has elapsed after the excitation, Meissner and Graffunder [56] finally came to the conclusion that the following method was most suitable. Both the emitting lamp and the absorption tube were operated with A.C. of the same frequency. The generators supplying the A.C. were so arranged that the phase of the alternating excitation of the emitting lamp could be made to lag behind that of the absorption tube by any amount. In this way, a beam of light from the lamp could be made to traverse the absorption tube at any desired time after the excitation of the absorption tube had ceased. For a detailed description of the circuits, the reader is referred to the paper of Meissner and Graffunder, and also to a more recent paper by Anderson[2].

The gas pressure in the emitting lamp is made quite small, and the current is kept as low as possible. The emitting layer is very small, almost a capillary. Under these conditions, it is wholly reasonable to assume that the frequency distribution of the emission line, arising from a transition from a high level to one of the metastable levels, is determined by the Doppler effect alone, uninfluenced by self-reversal. The absorption of this line by the metastable atoms in the absorption tube is measured by a photographic method, and is determined as a function of the time which elapses between the cessation of the excitation of the absorbing gas, and the passage of the absorbable light. The experimental results are expressed in the form of a "curve of decay", with absorption, A_1, plotted as ordinates, and time plotted as abscissas.

The next step is to translate the curve "A against t" into "n' against t", where n' is the average number of metastable atoms per c.c. responsible for the absorption. Although it is impossible to calculate n' in absolute magnitude, the quantity $k_0 l$, which contains n', can be calculated on the basis of the assumption that both the emission and the absorption lines are simple Doppler lines, or, if they show hyperfine structure, that they consist of a number of approximately equal, separate, Doppler components. Let k_0 be the absorption coefficient of the gas in the absorption tube at the centre of the line (or at the centre of each hyperfine-structure component), and l the length of the absorption tube. Then the absorption, A_1, is given by

$$A_1 = \frac{\int_{-\infty}^{\infty} e^{-\omega^2} (1 - e^{-k_0 l e^{-\omega^2}})\, d\omega}{\int_{-\infty}^{\infty} e^{-\omega^2}\, d\omega} \qquad \ldots\ldots(160).$$

The above expression will be recognized as the quantity A_α

defined in Chap. III, § 4d, with $\alpha = 1$. From the table of values of A_1 given in the Appendix, a graph may be drawn connecting $\log_{10} A_1$ and $\log_{10} (10 k_0 l)$. This curve is shown in Fig. 63, and enables one to read off $\log_{10} (10 k_0 l)$ (which is equal to $\log C n'$,

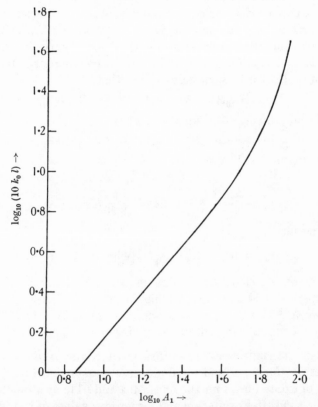

Fig. 63. Dependence of absorption on number of absorbing atoms.

where C is a constant) for any observed value of A_1. The experimental curve, "A_1 against t", can therefore be converted into a curve of "$\log_{10} C n'$ against t". It will be shown later that, when t is large enough,

$$n' = n_0 e^{-\beta t} \qquad \ldots\ldots(161)$$

or

$$\log_{10} n' = \log_{10} n_0 - \frac{\beta}{2 \cdot 30} t,$$

whence $$\log_{10} Cn' = \text{const.} - \frac{\beta}{2 \cdot 30}\, t \qquad \ldots \ldots (162),$$

which shows that the curve "$\log_{10} Cn'$ against t" should be a straight line, whose slope, multiplied by $2 \cdot 30$, is the exponential constant of decay of the metastable atoms.

The exponential constant, β, may be obtained directly from the original experimental curve of "A_1 against t", provided one uses only that portion of the experimental curve in which A_1 lies between $0 \cdot 3$ and $0 \cdot 8$. For, upon plotting $\log_{10}(10k_0 l)$ against A_1, it is immediately evident that

$$\log_{10}(10k_0 l) = 0 \cdot 3 + 1 \cdot 4 A_1 \quad [0 \cdot 3 < A_1 < 0 \cdot 8],$$

and again setting $10k_0 l$ equal to Cn',

$$\log_{10} n' = \text{const.} + 1 \cdot 4 A_1 \quad [0 \cdot 3 < A_1 < 0 \cdot 8].$$

Anticipating, as before,

$$n' = n_0 e^{-\beta t}$$

or $$\log_{10} n' = \log_{10} n_0 - \frac{\beta}{2 \cdot 30}\, t,$$

we get finally

$$A_1 = \text{const.} - \frac{\beta}{1 \cdot 4 \times 2 \cdot 3}\, t \quad [0 \cdot 3 < A_1 < 0 \cdot 8] \quad \ldots (163),$$

which shows that the original experimental curve of A_1 against t should be a straight line, in the region where A_1 lies between $0 \cdot 3$ and $0 \cdot 8$, whose slope, multiplied by $1 \cdot 4 \times 2 \cdot 3$, is the exponential constant of decay of the metastable atoms.

$13c$. EXPERIMENTAL RESULTS WITH NEON, ARGON, AND HELIUM. Meissner and Graffunder[56] measured the absorption of excited neon for the lines 6402 and 6143 as a function of the time after excitation and for various values of the neon pressure in the absorption tube. It can be seen from Fig. 64 that the absorption of these lines is determined by the number of $^3P_2$ atoms. As a matter of fact, all three ^3P states lie so close together that the total energy difference, $0 \cdot 09$ volt, is comparable to the average kinetic energy of the gas, kT, which, at 300° K., is $0 \cdot 026$ volt. Therefore, after the excitation has ceased, transitions occur so frequently among the three states that they may be regarded as one state. The experiments therefore may be considered to indicate the way in which atoms in all three states decrease in number after the excitation has ceased. From the experimental curves of A_1 against t, β was obtained in the manner described in § 13b. The values of β at various neon pressures are given in Table XLI. In Fig. 65 a

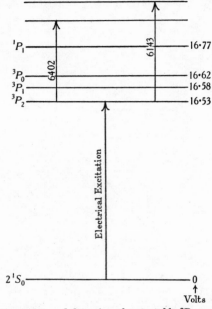

Fig. 64. Formation and detection of metastable $^3P_2$ neon atoms.

TABLE XLI

Neon pressure p, in mm.	Exponential constant β, in sec.$^{-1}$
0·24	6100
0·50	4400
1·02	3500
1·42	2800
2·15	3200
3·02	3900
5·60	5800

straight line is obtained when βp is plotted against p^2, showing that β depends on the pressure according to the relation

$$\beta = \frac{B}{p} + Cp \qquad \ldots\ldots(164),$$

where B and C are constants. Meissner and Graffunder's results yield the following values for B and C: $B = 2000$ and $C = 1100$.

With apparatus very similar to that used by Meissner and Graffunder, Anderson[2] studied the absorption of the argon line 7635 by excited argon as a function of the time after

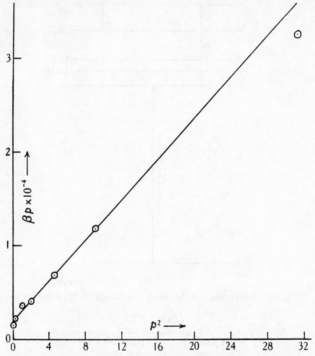

Fig. 65. Dependence of exponential constant on neon pressure.

excitation, at various argon pressures. It will be seen from Fig. 66 that the absorption of 7635 is an indication of the number of argon atoms in the metastable $^3P_2$ state. Unfortunately, Anderson did not give curves of absorption against time, and hence β cannot be calculated by the methods of §13b. The values of β, given in Table XLII, are therefore slightly inaccurate, but can still be used to obtain worth-while information. Plotting βp against p^2 for the temperature 300° K., all points except the last lie roughly on a straight line, indicating that Eq. (164) is approximately satisfied with $B = 160$ and $C = 120$. At the temperature 80° K., the curve of βp against p^2 is quite a good straight line with the constants $B = 15$ and

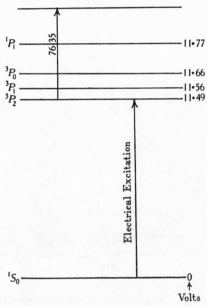

Fig. 66. Formation and detection of metastable $^3P_2$ argon atoms.

TABLE XLII

Temp. = 300° K.		Temp. = 80° K.	
Argon pressure p, in mm.	Exp. constant β	Argon pressure p, in mm.	Exp. constant β
0·215	815	0·050	330
0·42	408	0·080	183
0·61	347	0·125	120
0·694	283	0·23	97·6
0·805	315	0·36	84·5
1·0	660	0·605	142
—	—	1·0	183

$C = 170$. This line is shown in Fig. 67. The fact that Anderson's results agree well with Eq. (164) at the low temperature, and not so well at the high temperature, may be due to the fact that the three ^3P levels of argon are separated more than those

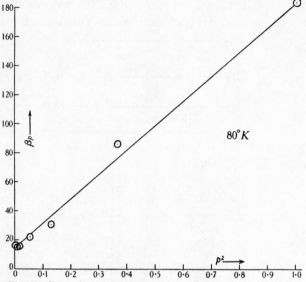

Fig. 67. Dependence of exponential constant on argon pressure.
(Temperature 80° K.)

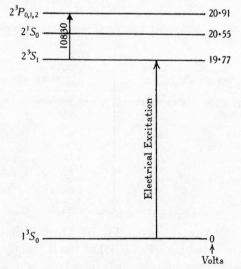

Fig. 68. Formation and detection of metastable $2\,^3S_1$ helium atoms.

of neon, and, at room temperature, they cannot be regarded as one composite level as in neon, but that transitions among the levels play a rôle which necessitates a much more complicated formula. At 80° K. these transitions are probably few in number, and hence the results express the behaviour of $^3P_2$ argon atoms alone.

With apparatus also similar to that of Meissner and Graffunder, Ebbinghaus [17] measured the absorption of the helium line 10830 by excited helium as a function of the time after the cut-off of the excitation. From Fig. 68 it is clear that the absorption of the 10830 line is an indication of the number of metastable $2\,^3S_1$ helium atoms. The experiments can be expected to yield information concerning the $2\,^3S_1$ state alone, since there are no other states that lie as close to it as in the

TABLE XLIII

Thickness of absorption tube a, in cm.	Helium pressure p, in mm.	Exp. constant β
1·65	2·5	810
1·65	3·7	760
3·0	3·7	603

case of neon and argon. Using the method of § 13 b to calculate β from the experimental curves of A_1 against t, the values given in Table XLIII were obtained. For reasons which will be given in the next section, β can also in this case be expected to obey Eq. (164), and the following values of B and C are obtained: when a is 1·65, $B = 1350$ and $C = 107$; when a is 3·0, $B = 770$ and $C = 107$.

13 d. THEORETICAL INTERPRETATION OF RESULTS WITH INERT GASES. The manner in which β varies with pressure gives an immediate clue as to the processes which metastable atoms perform in an excited gas after the excitation has been removed. The large values of β at low pressures indicate the rapidity with which metastable atoms diffuse to the walls, where they give up their energy. In this region an increase in pressure causes a decrease in diffusion rate. At higher pressures, the slowly rising values of β indicate a collision process

which either lowers the metastable atom to the normal state or raises it to a higher state of short life from which it radiates. The experiments that have just been described are capable therefore of yielding information concerning the diffusion and collision of metastable atoms. To obtain this information it is necessary to solve the following problem, which was first handled by Meissner and Graffunder[56] and later by Zemansky[113]. Consider a cylindrical tube of length l and radius c placed with its centre at the origin of cylindrical co-ordinates. Suppose that the tube is filled with a gas that has been electrically excited for a while and that a distribution of metastable atoms has been set up in the tube. Let the excitation be cut off, and let a cylindrical beam of light of radius b, which the metastable atoms are capable of absorbing, traverse the tube.

We shall make the following assumptions: (1) After the excitation has ceased, there is no further rate of formation of metastable atoms. (2) Metastable atoms diffuse to the walls where they lose their energy. (3) Metastable atoms perform impacts with normal atoms which raise the metastable atoms to a higher energy level. (4) The rate at which metastable atoms are being raised to a higher state by the absorption of the light is negligible compared to the rate at which (2) and (3) go on.

If n represent the number of metastable atoms per c.c., then, at any moment after the excitation has ceased, n will be given by

$$\frac{\partial n}{\partial t} + Zn = D\left(\frac{\partial^2 n}{\partial r^2} + \frac{1}{r}\frac{\partial n}{\partial r} + \frac{\partial^2 n}{\partial x^2}\right) \quad \ldots\ldots(165),$$

where D is the diffusion coefficient for metastable atoms, and Z is the number of impacts per sec. per metastable atom that are effective in raising the metastable atom to a higher radiating state. The boundary conditions are:

$$\text{when } t = 0, \quad n = f(r, x);$$

$$\text{when } t > 0, \quad n = 0 \begin{cases} \text{at } r = c, \\ \text{at } x = -l/2, \\ \text{at } x = +l/2. \end{cases}$$

The details of the solution of this problem will be found in the paper by Zemansky. The result is that, after a short time has elapsed, the average number n' of metastable atoms per c.c. in the path of the light beam is

$$n' = n_0 e^{-\beta t} \qquad \ldots\ldots(166),$$

where $$\beta = \left(\frac{5\cdot81}{c^2} + \frac{\pi^2}{l^2}\right)D + Z \qquad \ldots\ldots(167).$$

It will be seen that Eq. (166) is the result that was anticipated in §13b.

In the case of a rectangular absorption cell of infinite height, of thickness a, and length l, traversed by a beam of light travelling in the direction in which l is measured, the problem is very similar, and has been solved by Ebbinghaus[17]. The result is that, after a short time has elapsed, the average number n' of metastable atoms per c.c. in the path of the light beam is again an exponential function of the time, with an exponential constant, β, given by

$$\beta = \left(\frac{\pi^2}{a^2} + \frac{\pi^2}{l^2}\right)D + Z \qquad \ldots\ldots(168).$$

From kinetic theory, the diffusion coefficient is given by

$$D = \frac{4}{3\pi\sqrt{\pi}} \frac{\sqrt{\dfrac{RT}{M}}}{\sigma_d{}^2 N} \qquad \ldots\ldots(169),$$

where R is the universal gas constant, M is the molecular weight, N is the number of atoms per c.c. through which the metastable atoms are diffusing, and $\sigma_d{}^2$ the diffusion cross-section for a metastable atom. If p represent the pressure in mm., Eq. (169) becomes

$$D = 2\cdot25 \times 10^{-16} \frac{T\sqrt{\dfrac{T}{M}}}{\sigma_d{}^2} \cdot \frac{1}{p} \qquad \ldots\ldots(170).$$

Kinetic theory gives for the number of collisions capable of raising an atom to a state whose energy is ϵ volts higher, the expression

$$Z = 4\gamma\sigma^2 N \sqrt{\frac{\pi RT}{M}} \qquad \ldots\ldots(171),$$

where σ^2 is the cross-section for the process and γ is the fraction of all collisions of energy greater than ϵ. Two expressions have been used for γ in the past, namely

$$\gamma = e^{-\frac{\epsilon}{kT}} \qquad \ldots\ldots(172)$$

and

$$\gamma = \left(1 + \frac{\epsilon}{kT}\right) e^{-\frac{\epsilon}{kT}} \qquad \ldots\ldots(173),$$

but it is not yet clear which is to be preferred. Since, in all cases, ϵ is also not definitely known, it seems best to let the question remain open, and not to attempt to calculate σ^2 alone, but to allow the experiments to yield values of $\gamma\sigma^2$. Writing Eq. (171) in terms of pressure, p, in mm., we have

$$Z = 6\cdot30 \times 10^{23} \gamma\sigma^2 \sqrt{\frac{1}{MT}} \cdot p \qquad \ldots\ldots(174).$$

We are now in a position to give the empirical constants B and C a meaning in terms of $\sigma_d{}^2$ and $\gamma\sigma^2$. From Eqs. (167), (170) and (174), it is clear that β obeys the equation given empirically in § 13c, namely,

$$\beta = \frac{B}{p} + Cp,$$

and, in the case of a cylindrical absorption tube,

$$B = \left(\frac{5\cdot81}{c^2} + \frac{\pi^2}{l^2}\right) \times 2\cdot25 \times 10^{-16} \frac{T\sqrt{\frac{T}{M}}}{\sigma_d{}^2} \ldots\ldots(175),$$

and for the case of a rectangular absorption tube,

$$B = \left(\frac{\pi^2}{a^2} + \frac{\pi^2}{l^2}\right) \times 2\cdot25 \times 10^{-16} \frac{T\sqrt{\frac{T}{M}}}{\sigma_d{}^2} \qquad \ldots\ldots(176),$$

and in both cases,

$$C = 6\cdot30 \times 10^{23} \gamma\sigma^2 \sqrt{\frac{1}{MT}} \qquad \ldots\ldots(177).$$

In Table XLIV all the experimental quantities are given which, with the aid of the above equations, enable us to calculate $\sigma_d{}^2$ and $\gamma\sigma^2$.

TABLE XLIV

Gas	Molecular weight M	Temp. T, ° K.	Radius c, in cm.	Thickness a, in cm.	Length l, in cm.	B	C
Ne	20·2	300	1·8	—	12	2000	1100
A	39·9	300	2·5	—	Large	160	120
A	39·9	80	2·5	—	Large	15	170
He	4	300	—	1·65	5·2	1350	107
He	4	300	—	3·0	5·2	770	107

The final results are given in Table XLV.

TABLE XLV

Metastable atom	Conditions	$\sigma_d{}^2$ cm.$^2 \times 10^{16}$	$\gamma\sigma^2$ cm.$^2 \times 10^{16}$
Ne ($^3P_{2,1,0}$)	—	2·44	·00136
A ($^3P_2$)	$T = 300°$ K.	10·8	·000208
A ($^3P_2$)	$T = 80°$ K.	15·9	·000153
He ($2\,^3S_1$)	$a = 1·65$	17·3	·000059
He ($2\,^3S_1$)	$a = 3·0$	11·1	·000059

Not very much that is definite can be said about the values of $\sigma_d{}^2$ and $\gamma\sigma^2$ in Table XLV. In the first place, the theory which enabled these quantities to be calculated has one serious deficiency. It was assumed that, after the removal of the excitation, no further rate of formation of metastable atoms took place. This means that all sorts of collision processes that might be present were neglected. It is conceivable, for example, that metastable atoms could be formed by recombination of ions and electrons, or by collisions of atoms in higher states with either normal atoms or electrons. If such processes were taken into account, the exponential constant would not be represented by the simple Eq. (164) and would involve so many unknown quantities as to be useless. Furthermore, if collisions with electrons play an important rôle, the electron temperature would have to be considered, since the experiments of Kopfermann and Ladenburg [39] and Mohler [68] indicate that this temperature determines the number of atoms in the different excited states when the electron concentration is high, and, therefore, would affect the results even at moderate electron concentrations. An attempt was made by Anderson[1] to formulate a theory, taking into account collision processes other than the one considered in the present theory, but with little success. The mathematical complications are great, and too many unjustified simplifying assumptions had to be made.

If the present theory is at all justified, the values of $\sigma_d{}^2$ can be regarded as satisfactory, except in the case of neon, where it is unexpectedly low. The values of $\gamma\sigma^2$ can also be regarded as sensible with the exception of helium, where it is much too large, considering the fact that the metastable helium atom must be raised to the $2\,{}^1S_0$ state, which is $0\cdot78$ volt higher. On the whole, neither the experimental nor the theoretical parts of this field have yet been developed to a point where they are capable of yielding very reliable information.

13e. METHODS OF STUDYING METASTABLE MERCURY ATOMS IN NITROGEN. To produce metastable $6\,{}^3P_0$ mercury atoms in the presence of nitrogen, it is merely necessary to excite the mercury atoms to the $6\,{}^3P_1$ state by illuminating with 2537, and then rely upon collisions with the nitrogen molecules to bring them to the $6\,{}^3P_0$ state. The number of metastable atoms can then be measured in various ways, of which three have been used. Pool[77] illuminated a quartz cell containing a mixture of mercury vapour and nitrogen with the whole arc spectrum from a water-cooled mercury arc, and then, with a rotating wheel, cut this light off. A moment later, a beam of light of wave-length 4047 was allowed to traverse the tube. By a photographic method, the absorption of the 4047 line was measured as a function of the time that elapsed after the excitation ceased. From Fig. 1 it is clear that the 4047 line is absorbed only by $6\,{}^3P_0$ atoms, and therefore a curve of absorption against time is an indication of the decay of $6\,{}^3P_0$ atoms. A rough determination of the exponential constant of decay at various nitrogen pressures[113] yielded a curve of β against p which had the same characteristics as that of neon and argon, i.e. as the nitrogen pressure was increased, β first decreased to a minimum, and then increased. More recently, Pool[78] repeated these experiments under more advantageous conditions, and found that the curves of absorption against time showed an anomalous behaviour, which he interpreted as being due to long-lived metastable nitrogen molecules. This interpretation is somewhat in doubt, for the reason that the method of translating the values representing the absorption of 4047 into the number of metastable atoms is inaccurate. To understand this point, it must be emphasized that the lamp which emitted the 4047 line was not like the emission lamps used by Meissner and Graffunder, Anderson, and Ebbinghaus. It was not constructed and operated so as to emit a pure Doppler line, but instead, under conditions in which one is entitled to expect both broadening and self-reversal of the 4047 line. The relation, therefore, between absorption and the number of absorbing atoms was presumably quite different from the simple exponential one assumed by Pool. The anomalous character of the absorption-time curves is, in the opinion of the authors, to be attributed to this cause, rather than to the presence of metastable nitrogen molecules.

A very ingenious method of measuring the decrease in the number of metastable mercury atoms present in a mixture of mercury vapour and nitrogen, after the removal of the excitation, was used by Asada, Ladenburg and Tietze[3]. They allowed the metastable mercury atoms to absorb the line 4047, and at the same time measured the intensity of the green line 5461. It is clear from Fig. 1 that the intensity of the 5461 line depends on the number of $7\,^3S_1$ mercury atoms, which in turn depends on the amount of absorption of 4047. Asada[4] used this method to study the decay of metastable mercury atoms in nitrogen, but, unfortunately, did not give enough data to enable a calculation of σ_d^2 and $\gamma\sigma^2$ to be made.

Webb and Messenger[102] and, at about the same time, Samson[84] studied the same problem by still a third method. They relied on collisions between $6\,^3P_1$ mercury atoms and nitrogen to produce metastables, and collisions between metastables and nitrogen to produce $6\,^3P_1$ atoms again. The radiation from these $6\,^3P_1$ atoms, after the optical excitation had ceased, was used as an indication of the number of $6\,^3P_0$ atoms. The experiment was carried out in a very simple manner. The light

from a cooled mercury arc was sent through a slit past which
a toothed wheel rotated at high speed. In the time interval
when the slit was not covered, this light fell on a quartz cell
containing mercury vapour and nitrogen. During the time
interval when the slit was covered, a small hole in another
wheel rotated between the other face of the cell and the colli-
mator of a spectrograph. The radiation emitted by the cell
during this time interval caused a circular trace on a photographic plate, which represented the decay of the radiation.
The decay was found to be exponential after about 10^{-4} sec.
had elapsed, and the exponential constant was measured at
various nitrogen pressures. The mercury vapour pressure
remained constant during the experiment at a value corresponding to a temperature of 28° C.

13f. RESULTS AND INTERPRETATION WITH METASTABLE
MERCURY ATOMS IN NITROGEN. The theory of this method was
worked out in great detail by Samson. On account of its com-
plexity, only the salient features can be given here. First of
all, both $6\,^3P_0$ and $6\,^3P_1$ atoms were considered to diffuse
through the nitrogen with the same diffusion coefficient, and
to be destroyed at the walls. Second, the diffusion of the 2537
radiation that was imprisoned in the mercury vapour was
taken into account by calculating the rate at which this
radiation would leave the mercury vapour if no nitrogen were
present. This involved the calculation of an equivalent absorption coefficient [see § 8c] and the use of Milne's theory [§ 12b].
Finally, collision processes were considered in which nitrogen
molecules produced the following transitions in $6\,^3P_0$ and $6\,^3P_1$
atoms: (1) $6\,^3P_0 \to 6\,^3P_1$, (2) $6\,^3P_1 \to 6\,^3P_0$, (3) $6\,^3P_1 \to 6\,^1S_0$, and
(4) $6\,^3P_0 \to 6\,^1S_0$. On the basis of these ideas, Samson obtained
the result that the exponential constant, β, should depend
upon the nitrogen pressure according to the relation

$$\beta = \frac{X}{p} + \frac{Yp + Wp^2}{p + Z} \qquad \ldots \ldots (178),$$

where W, X, Y and Z are constants at a given temperature.
The experimental curves, which were obtained for three
different temperatures, 301° K., 374° K. and 486° K., were

found to fit Eq. (178) very satisfactorily, enabling the values of the constants to be obtained. From these constants the following results were obtained: the diffusion cross-section, $\sigma_d{}^2$, increased very slowly with temperature, being $15 \cdot 5 \times 10^{-16}$, $17 \cdot 7 \times 10^{-16}$ and $18 \cdot 4 \times 10^{-16}$ at the temperatures $301° K.$, $374° K.$ and $486° K.$ respectively. The other cross-sections remained independent of temperature and were for

$$6\,{}^3P_0 \rightarrow 6\,{}^3P_1, \quad \sigma_0{}^2 = 6 \cdot 7 \times 10^{-18}\,\text{cm.}^2,$$

$$6\,{}^3P_1 \rightarrow 6\,{}^3P_0, \quad \sigma_1{}^2 = 3 \cdot 1 \times 10^{-17}\,\text{cm.}^2,$$

$$6\,{}^3P_1 \rightarrow 6\,{}^1S_0, \quad \Sigma_1{}^2 \leqq 2 \cdot 2 \times 10^{-18}\,\text{cm.}^2,$$

$$6\,{}^3P_0 \rightarrow 6\,{}^1S_0, \quad \Sigma_0{}^2 \leqq 2 \cdot 0 \times 10^{-22}\,\text{cm.}^2.$$

The value of $\sigma_1{}^2$, $3 \cdot 1 \times 10^{-17}$ cm.^{2}, can be compared with Zemansky's value of the same quantity [see Table XXIV], obtained by measuring the quenching of mercury resonance radiation by nitrogen, namely, $2 \cdot 74 \times 10^{-17}$ cm.^{2}. The agreement is quite satisfactory.

13g. METASTABLE MERCURY ATOMS IN MERCURY VAPOUR. Webb[100] showed that, when a metastable mercury atom strikes a metal plate, an electron is liberated. Experiments of Oliphant[75] on metastable helium atoms, and quite recent experiments of Sonkin[91] on metastable mercury atoms, confirm this result. In Webb's experiments, the electrons liberated from a plate by metastable mercury atoms were drawn to a positive grid, and the resulting current was used as a measure of the number of metastable atoms striking the plate. The metastable atoms were produced in another part of the tube by electrons liberated from a hot cathode and accelerated by a grid near by. By applying an alternating accelerating potential, and another alternating potential between the plate and its grid, and by varying the frequency of these potentials, Webb was able to measure the rate at which metastable mercury atoms arrived at the plate after diffusing through mercury vapour. This same experiment was carried out in a more refined manner by Coulliette[12], who was able to calculate from his experiments the effective diffusion cross-section of a metastable mercury atom. The result was 20×10^{-16} cm.^{2}.

Coulliette's experiments also indicated that collisions between metastable and normal mercury atoms occur which destroy the metastable atom (probably by raising it to the $6\,^3P_1$ state, from which it radiates). The value of $\gamma\sigma^2$ for this process was found to be $0\cdot016 \times 10^{-16}$ cm.^{2}.

13h. THE SIMULTANEOUS PRODUCTION AND DESTRUCTION OF METASTABLE ATOMS. During an arc discharge in an inert gas, or during the optical excitation of a mixture of mercury vapour and nitrogen, metastable atoms are being formed, are diffusing and are performing collisions, all at the same time. The situation is a steady state, where, in any unit volume, the rate at which metastable atoms are forming, due to excitation and diffusion, is equal to the rate at which they are being destroyed by collision. Many experiments have been performed on metastable atoms in the steady state, but they are not suited for quantitative treatment because of the lack of knowledge of the rate at which the metastable atoms are being formed. It is clear that, to calculate this rate, it would be necessary to know the electron concentration, the electron excitation function, and the number of transitions from higher states, in the case of an arc discharge; and, in the case of optical excitation, the intensity of the exciting light, absorption coefficient, etc. The advantage of experiments of the kind described in the preceding sections is that most of these processes, in the after-glow, do not exist or can be ignored. Chief among the experiments on the steady state should be mentioned those of Eckstein [18] on neon and mixtures of neon with foreign gases. These experiments indicate that metastable neon atoms are destroyed by impact with foreign gas molecules, hydrogen being the most effective, then nitrogen, and helium the least. An attempt at a quantitative treatment of Eckstein's experiments is to be found in a paper by Zemansky [113].

The measurements of Kopfermann and Ladenburg [39] of the number of metastable neon atoms in the positive column of a neon arc indicate that, at low current densities, metastable neon atoms are destroyed by collisions with normal neon atoms, and at high current densities, impacts of the second kind take place with electrons. As a result of collisions of the first and of the second kind with electrons, at high current densities, neon atoms are distributed among the various excited states according to the Boltzmann equation, in which the temperature is the *electron temperature*.

The experiments of Klumb and Pringsheim[38] on the absorption by metastable mercury atoms of the 4047 line at various foreign gas pressures indicate very graphically the various collision processes that a metastable mercury atom can perform. Finally the works of Found and Langmuir[20], Kenty[35], and many others, show that metastable atoms may play a very important rôle in the maintenance of an arc discharge in an inert gas.

REFERENCES TO CHAPTER IV

[1] Anderson, J. M., *Can. Journ. Res.* 2, 13 (1930).
[2] —— *ibid.* 4, 312 (1931).
[3] Asada, T., Ladenburg, R. and Tietze, W., *Phys. Zeits.* 29, 549 (1928).
[4] Asada, T., *ibid.* 29, 708 (1928).
[5] Bates, J. R., *Proc. Nat. Acad. Sci.* 14, 849 (1928).
[6] —— *Journ. Amer. Chem. Soc.* 52, 3825 (1930).
[7] —— *ibid.* 54, 569 (1932).
[8] Bender, P., *Phys. Rev.* 36, 1535 (1930).
[9] Bonhoeffer, K. F., *Z. f. Phys. Chem.* 116, 391 (1925).
[10] Bonner, T. W., *Phys. Rev.* 40, 105 (1932).
[11] Child, C. D., *ibid.* 38, 699 (1931).
[12] Coulliette, J. H., *ibid.* 32, 636 (1928).
[13] Dennison, D. M., *ibid.* 31, 503 (1928).
[14] Dorgelo, H. B., *Physica*, 5, 429 (1925).
[15] Dorgelo, H. B. and Washington, T. P. K., *Ak. Wet. Amst.* 35, 1009 (1926).
[16] Duffendack, O. S. and Thomson, K., *Phys. Rev.* 43, 106 (1933).
[17] Ebbinghaus, E., *Ann. d. Phys.* 7, 267 (1930).
[18] Eckstein, L., *ibid.* 87, 1003 (1928).
[19] Foote, P. D., *Phys. Rev.* 30, 288 (1927).
[20] Found, C. G. and Langmuir, I., *ibid.* 39, 237 (1932).
[21] Frenkel, J., *Z. f. Phys.* 59, 198 (1930).
[22] Füchtbauer, C., Joos, G. and Dinkelacker, O., *Ann. d. Phys.* 71, 204 (1923).
[23] Garrett, P. H., *Phys. Rev.* 40, 779 (1932).
[24] Gaviola, E., *ibid.* 33, 309 (1929); 34, 1049 (1929).
[25] Gouy, G. L., *Ann. Chim. Phys.* 18, 5 (1879).
[26] v. Hamos, L., *Z. f. Phys.* 74, 379 (1932).
[27] Hayner, L. J., *Phys. Rev.* 26, 364 (1925).

[28] v. d. Held, E. F. M. and Ornstein, S., *Z. f. Phys.* **77**, 459 (1932).

[29] Hogness, T. R. and Franck, J., *ibid.* **44**, 26 (1927).

[30] Holtsmark, J., *ibid.* **34**, 722 (1925).

[31] Jablonski, A., *ibid.* **70**, 723 (1931).

[32] Kallmann, H. and London, F., *Z. f. Phys. Chem.* B **2**, 207 (1929).

[33] Kaplan, J., *Phys. Rev.* **31**, 997 (1928).

[34] Kenty, C., *ibid.* **42**, 823 (1932).

[35] —— *ibid.* **43**, 181 (1933).

[36] Kisilbasch, B., Kondratjew, V. and Leipunsky, A., *Sow. Phys.* **2**, 201 (1932).

[37] Klein, O. and Rosseland, S., *Z. f. Phys.* **4**, 46 (1921).

[38] Klumb, H. and Pringsheim, P., *ibid.* **52**, 610 (1928).

[39] Kopfermann, H. and Ladenburg, R., *Naturwiss.* **19**, 513 (1931).

[40] Kuhn, H. and Oldenberg, O., *Phys. Rev.* **41**, 72 (1932).

[41] Kulp, M., *Z. f. Phys.* **79**, 495 (1932).

[42] Kunze, P., *Ann. d. Phys.* **8**, 500 (1931).

[43] Landau, L., *Sow. Phys.* **1**, 89 (1932).

[44] —— *ibid.* **2**, 46 (1932).

[45] Lawrence, E. O. and Edlefsen, N. E., *Phys. Rev.* **34**, 233 (1929).

[46] Lenz, W., *Z. f. Phys.* **25**, 299 (1924).

[47] —— *ibid.* **80**, 423 (1933).

[48] Locher, G. L., *Phys. Rev.* **31**, 466 (1928).

[49] Lochte-Holtgreven, W., *Z. f. Phys.* **47**, 362 (1928).

[50] London, F., *ibid.* **74**, 143 (1932).

[51] Lorentz, H. A., *Proc. Amst. Acad.* **18**, 134 (1915).

[52] Mannkopff, R., *Z. f. Phys.* **36**, 315 (1926).

[53] Margenau, H., *Phys. Rev.* **40**, 387 (1932).

[54] — *ibid.* **43**, 129 (1933).

[54a] Margenau, H. and Watson, W. W., *ibid.* **44**, 92 (1933).

[55] Meissner, K. W., *Phys. Zeits.* **26**, 687 (1925).

[56] Meissner, K. W. and Graffunder, W., *Ann. d. Phys.* **84**, 1009 (1927).

[57] Mensing, L., *Z. f. Phys.* **34**, 611 (1925).

[58] Michelson, A., *Astrophys. Journ.* **2**, 251 (1895).

[59] Milne, E. A., *Journ. Lond. Math. Soc.* **1**, 1 (1926).

[60] Minkowski, R., *Z. f. Phys.* **36**, 839 (1926).

[61] — *ibid.* **55**, 16 (1929).

[62] Mitchell, A. C. G., *ibid.* **49**, 228 (1928).

[63] —— *Journ. Frankl. Inst.* **206**, 817 (1928).

[64] Mohler, F. L., Foote, P. D. and Chenault, R. L., *Phys. Rev.* **27**, 37 (1926).

[65] Mohler, F. L., *ibid.* **29**, 419 (1927).

[66] Mohler, F. L. and Boeckner, C., *Bureau of Stand. Journ. Res.* **5**, 51 (1930).

[67] —— —— *ibid.* **5**, 399 (1930).

[68] Mohler, F. L., *ibid.* **9**, 493 (1932).

[69] Morse, P. M. and Stueckelberg, E. C. G., *Ann. d. Phys.* **9**, 579 (1931).

[70] Morse, P. M., *Rev. Mod. Phys.* **4**, 577 (1932).

[71] Neumann, E. A., *Z. f. Phys.* **62**, 368 (1930).

[72] Noyes, W. A., Jr., *Journ. Amer. Chem. Soc.* **49**, 3100 (1927).

[73] Noyes, W. A., Jr., *Journ. Amer. Chem. Soc.* **53**, 514 (1931).
[74] Oldenberg, O., *Z. f. Phys.* **51**, 605 (1928).
[75] Oliphant, M. L. E., *Proc. Roy. Soc.* A **124**, 228 (1929).
[76] Orthmann, W., *Ann. d. Phys.* **78**, 601 (1925).
[77] Pool, M. L., *Phys. Rev.* **33**, 22 (1929).
[78] —— *ibid.* **38**, 955 (1931).
[79] Prileshajewa, N., *Sow. Phys.* **2**, 351 (1932).
[80] —— *ibid.* **2**, 367 (1932).
[81] Reiche, F., *Verh. d. D. Phys. Ges.* **15**, 3 (1913).
[82] Rice, O. K., *Proc. Nat. Acad. Sci.* **17**, 34 (1931).
[83] —— *Phys. Rev.* **38**, 1943 (1931).
[84] Samson, E. W., *ibid.* **40**, 940 (1932).
[85] Schönrock, O., *Ann. d. Phys.* **20**, 995 (1906).
[86] Schuster, A., *Astrophys. Journ.* **21**, 1 (1905).
[87] Schütz, W., *Z. f. Phys.* **45**, 30 (1927).
[88] —— *ibid.* **71**, 301 (1931).
[89] Schütz-Mensing, L., *ibid.* **61**, 655 (1930).
[90] Senftleben, H., *Ann. d. Phys.* **47**, 949 (1915).
[91] Sonkin, S., *Phys. Rev.* **43**, 788 (1933).
[92] Stern, O. and Volmer, M., *Phys. Zeits.* **20**, 183 (1919).
[93] Stuart, H., *Z. f. Phys.* **32**, 262 (1925).
[94] Stueckelberg, E. C. G., *Helv. Phys. Acta*, **5**, 370 (1932).
[95] Terenin, A., *Z. f. Phys.* **37**, 98 (1926).
[96] Terenin, A. and Prileshajewa, N., *Z. f. Phys. Chem.* B **13**, 72 (1931).
[97] —— —— *Sow. Phys.* **2**, 337 (1932).
[98] Trumpy, B., *Intensität und Breite der Spektrallinien*, Trondhjem (1927).
[99] Voigt, W., *Münch. Ber.* p. 603 (1912).
[100] Webb, H. W., *Phys. Rev.* **24**, 113 (1924).
[101] Webb, H. W. and Messenger, H. A., *ibid.* **33**, 319 (1929).
[102] —— —— *ibid.* **40**, 466 (1932).
[103] Weingeroff, M., *Z. f. Phys.* **67**, 679 (1931).
[104] Weisskopf, V., *ibid.* **75**, 287 (1932).
[105] —— *Phys. Zeits.* **34**, 1 (1933).
[106] Wilson, H. A., *Phil. Trans. Roy. Soc.* **216**, 63 (1916).
[107] Winans, J. G., *Z. f. Phys.* **60**, 631 (1930).
[108] Wood, R. W., *Phys. Zeits.* **13**, 353 (1912).
[109] —— *Phil. Mag.* **27**, 1018 (1914).
[110] Wood, R. W. and Mohler, F. L., *Phys. Rev.* **11**, 70 (1918).
[111] Zemansky, M. W., *ibid.* **29**, 513 (1927).
[112] —— *ibid.* **31**, 812 (1928).
[113] —— *ibid.* **34**, 213 (1929).
[114] —— *ibid.* **36**, 219 (1930).
[115] —— *ibid.* **36**, 919 (1930).
[116] —— *ibid.* **42**, 843 (1932).
[117] Zener, C., *ibid.* **38**, 277 (1931).

CHAPTER V

THE POLARIZATION OF RESONANCE RADIATION

1. INTRODUCTION

It has long been known that the band fluorescence of sodium and iodine vapours is polarized if observed in a direction at right angles to the exciting light beam, but it was not until 1922 that Rayleigh [40] discovered that the 2537 line of mercury was polarized if excited as resonance radiation by a polarized light source. This effect was investigated more completely by Wood [53] and by Wood and Ellett [54]. They observed that if mercury vapour, at low pressure, is excited by polarized light from a quartz mercury arc, then (in zero magnetic field) the re-emitted resonance line is polarized with its electric vector in the same direction as that of the exciting light. In the absence of any magnetic field the resonance radiation was almost completely linearly polarized, whereas in the presence of small magnetic fields in certain directions the polarization was found to decrease. The addition of foreign gases was also found to diminish the degree of polarization. On the other hand, experiments on the polarization of sodium resonance radiation, consisting of the two D lines, showed that the D_2 line was about 20 per cent. polarized and the other completely unpolarized under all circumstances. To explain these difficulties, it will be well to start with the case of mercury and discuss some further experiments by Hanle [20] in the light of the classical theory and also on the Bohr theory. The modern quantum-mechanical theory can be shown to be in accord with the Bohr theory.

2. GENERAL DESCRIPTION OF APPARATUS FOR POLARIZATION WORK

Before discussing the various experiments which have been performed to show the polarization of resonance radiation, it will be necessary to describe the essential apparatus used. The arrangement of apparatus in the several experiments is somewhat varied, but consists essentially of a light source, polarizer, resonance tube, analyser and spectrograph or photocell.

In general, measurements on polarization of resonance radiation are made by observing the resonance radiation coming off from a resonance tube in a direction perpendicular to the beam of exciting radiation, as is shown in Fig. 69. Radiation from a source S is passed through a lens L_1, and Nicol prism N_1, to polarize it, and is converged on the resonance tube T. In all polarization work the angular aperture, α, of the exciting beam should be kept as small as possible. The reason for this is

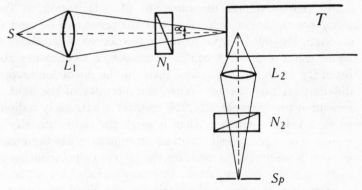

Fig. 69. Apparatus for studying polarization of resonance radiation.

apparent, since the electric vectors of any ray of the beam are at right angles to that ray. Thus, if observations are to be made in a direction perpendicular to the direction of the exciting beam when the primary light is polarized with its electric vector perpendicular to the plane of Fig. 69, and is falling on the resonance tube under an angular aperture α, the primary light cannot be said to be 100 per cent. polarized, since its electric vectors have a maximum deviation $\alpha/2$ from the plane of polarization. A method of correcting for this effect has been given by Gaviola and Pringsheim[16] and Heydenburg[25].

In case the activating wave-length of the primary beam lies in the ultra-violet, a Nicol prism cannot be used, since the Canada balsam cement in such prisms absorbs all light of wavelength below 3200. In this case a Glans prism of quartz, cemented together with glycerine, may be used for wave-lengths down to about 2000. In order to use this type of prism the light must pass through it parallel or under an angular aperture less than 6°. Otherwise the use of the Glans prism is similar to that of a Nicol. A calcite block, which transmits well down to 2000, may be also used as polarizer. When this is employed, the convergent light from the lens L_1 of Fig. 69 passes through the block and two images of the source, polarized at right angles to each other, are formed on the resonance tube. One of these images is usually screened off, thus giving a polarized beam.

For detection and measurement of polarization of the resonance radiation, a Nicol or a Glans prism can be used as shown in Fig. 69. In order to obtain the degree of polarization when using a Nicol prism it is necessary to measure the intensity of the light passing through the Nicol for several different settings thereof. When the intensity of the light is measured photographically, the method is extremely tedious so that usually, when a Nicol is used, the light intensity is recorded on a photo-cell. Such an arrangement has been used by von Keussler [49] to measure the degree of polarization of mercury resonance radiation. One may make a plot of photo-electric current against the setting of the Nicol (in degrees), from which the degree of polarization can be obtained by measuring the height of the maxima of the curve and comparing them with a light source of the same intensity which is known to be fully polarized.

When photographic measurements of intensity are to be made, using a spectrograph for example, it is found convenient to employ a double-image prism of the Wollaston or Rochon type. If light from the resonance tube is made to converge through the prism on to the slit of a spectrograph two images of the line or lines emitted in the resonance tube are seen on the photographic plate, the two images being polarized at right angles to each other. By measuring the relative intensity of these two images the degree of polarization of the light may be calculated. In using this method a certain amount of pre-caution must be taken, since the loss of light in the spectrograph due to reflection from the faces of the dispersing prism is dependent upon the polarization of the light striking it, which may easily falsify the results. When using the double-image prism the light leaving the prism must be depolarized, or calibration experiments must be made. A special double-image prism has been described by Hanle in which the two images are depolarized after leaving the prism.

A more exact means of measuring the polarization is by the method of Cornu. In this method the light to be investigated is made parallel and sent through two Wollaston prisms. If partially polarized light is incident on the apparatus, four images will, in general, be formed. Suppose the two images formed by the first prism are polarized parallel to X and Y, respectively, and that the second prism makes an angle α with the first. Of the four images formed by the second prism, two will be polarized parallel to x and two parallel to y, where the angle (X, x) is α. The intensity of the four images will then be

$$J_{Xx} = I_X \cos^2 \alpha; \quad J_{Xy} = I_X \sin^2 \alpha;$$
$$J_{Yx} = I_Y \sin^2 \alpha; \quad J_{Yy} = I_Y \cos^2 \alpha;$$

where I_X and I_Y are the intensities of the original radiation polarized parallel to X and Y, respectively. The procedure is to find the value of α for which $J_{Xx} = J_{Yx}$ or $J_{Xy} = J_{Yy}$. At this value of α, the polarization is given by

$$P = \frac{I_X - I_Y}{I_X + I_Y} = \pm \cos 2\alpha,$$

depending on which images are compared. If a Glans prism is used instead of the second Wollaston, two images are formed and a similar relation between the intensity of the images exists. The advantage of this method is twofold: (1) it is easy to find the setting of the prism for which two images are equal, and (2) there is no correction to be made to the polarization for loss of light due to reflection, since both Glans and Wollaston prisms are cut in such a way that the incident light traverses the prism perpendicular to its face.

Another means of detecting polarized light, and this is especially good for detecting a small degree of polarization, is the Savart plate used in conjunction with a Nicol prism. If plane polarized white light passes through a Savart plate and then through a Nicol prism, coloured fringes are seen for certain positions of the Nicol prism. If the light is analysed by a spectrograph, the apparatus can be so arranged that each spectral line is crossed by light and dark fringes. The distinctness of these fringes gives the degree of polarization. The actual amount of polarization is usually obtained by placing a number of glass plates between the polarized light source and the Savart plate. The plates are rotated about an axis until the fringes formed by the Savart plate disappear, indicating that the polarization of the original light has been compensated. From the angle of rotation of the plates and their index of refraction, the degree of polarization can be calculated. If two lines are observed which are polarized at right angles to each other, the maxima of the fringes of the one line come at about the same place as the minima of those of the other line, if the wave-lengths of the lines are not very different.

In order to measure changes in the angle of polarization of resonance radiation a system of quartz wedges or a Babinet compensator may be used. The angle of rotation is obtained by measuring the shift of the position of the fringes formed in the system of wedges.

It is hardly necessary to remark that when lenses are used between the resonance tube and the apparatus for detecting polarization they should be non-rotatory. In the ultra-violet region a fused quartz lens or a matched pair of crystalline quartz lenses of left- and right-handed rotation should be used.

3. HANLE'S EXPERIMENTS ON MERCURY VAPOUR

Hanle[20] made a thorough study of the polarization of the mercury resonance line 2537. For this investigation he used a Glans prism as polarizer and a Savart plate arrangement as analyser. The apparatus was arranged in such a way that the exciting light was incident on a resonance tube in the Z direction (Fig. 70) and the resonance radiation is observed along OY, with Savart plate, Nicol prism and a photographic plate. The resonance tube was placed in a system of coils in such a way that the earth's magnetic field was always compensated and magnetic fields of known strengths in given directions could be supplied. The pressure of mercury in the tube was 10^{-4} mm.

If the exciting radiation is polarized with its electric vector in the X direction and there is no magnetic field on the tube, the resonance radiation is found to be highly polarized (about 90 per cent.) in the X direction. On the other hand, if the exciting light is polarized along Y, the resonance radiation is unpolarized and its intensity extremely weak. If the direction of polarization of the exciting light is changed slowly from Y to X, the polarization and the intensity of the resonance radiation increase.

If the direction of the electric vector of the exciting light is kept constant and parallel to X, and a magnetic field (about

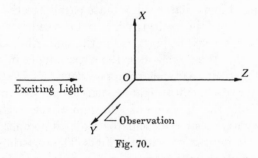

Fig. 70.

25 to 100 gauss) placed in the X direction, the polarization of the resonance radiation remains unchanged, that is, parallel to X. With the field parallel to Z the polarization of the resonance radiation is again high and parallel to X. If, however, the field is parallel to Y, that is along the direction of observation, the resonance radiation is completely unpolarized but is strong. Suppose the field in the direction of Y is not strong but weak and can be varied from zero to a few gauss. With zero field the resonance radiation is, of course, 90 per cent. polarized in the X direction. On increasing the field the degree of polarization is found to decrease and, for small fields, its direction is changed slightly from the X direction. As the field increases still further the degree of polarization diminishes to zero.

Finally, if the electric vector of the exciting light is parallel to Y, and there is a strong field parallel to X, the resonance radiation is highly polarized parallel to Z, and on rotating the field from X to Z, the direction of polarization rotates from Z to X, being always perpendicular to the field and keeping its degree of polarization constant.

Hanle also found that, on using circularly or elliptically polarized exciting light, and observing at an angle of $20°$ to the incident beam, the resonance radiation was circularly or elliptically polarized in the same manner as the incident beam.

4. THEORY OF HANLE'S EXPERIMENTS

4*a*. CLASSICAL THEORY. It is obvious that the classical theory will roughly explain all the results if one considers the series electron of the mercury atom to act as a classical oscillator. Thus, the oscillator will vibrate parallel to the direction of polarization of the exciting light, and the radiation emitted by the oscillator will be polarized in the same direction as the exciting light, thus explaining the experiments in zero field with the incident beam polarized parallel to X. In the experiments where the incident light is polarized parallel to Y, one is looking along the direction of vibration of the oscillator and the theory says that the oscillator radiates no energy in this direction, in agreement with the facts. The experiments with various orientations of magnetic field are also explained on the classical theory when one remembers that the electron will precess about a magnetic field giving rise to circularly polarized light when viewed along the field (classical Zeeman effect) or, when viewed perpendicular to the plane of the field, to linearly polarized light (perpendicular to the field), since only the simple harmonic components of the circular vibration are seen. Thus, when the electric vector of the exciting light is parallel to X, and there is a strong field parallel to Z, the electron of the classical model will precess about the Z axis and the light observed in the direction Y will appear polarized parallel to X, since only the simple harmonic components of the circular vibration are seen. The case in which the plane of polarization of the emitted light rotates, when the direction of the magnetic field is rotated from X to Z, is also easily explained by these considerations.

In order to explain the fact that, when the resonance radiation is observed in the direction of the magnetic field, it becomes depolarized with increasing field, it is sufficient to assume the classical model to be a damped oscillator. If the oscillator is excited by light polarized in the X direction it will start to vibrate parallel to the X axis but will precess about the

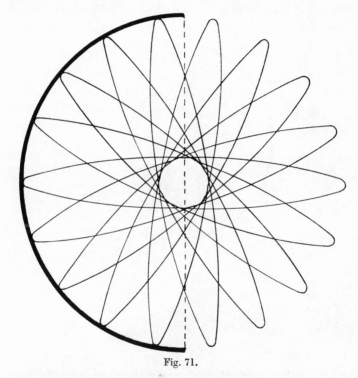

Fig. 71.

field, its amplitude of oscillation dying down with the time due to damping. The path described by the oscillator when viewed along the field will take the form of a rosette. If the precession velocity is large compared to the damping (that is, large magnetic field), the rosette will be symmetrical, as shown in Fig. 71. In this case, since the rosette is perfectly symmetrical, it is clear that the light from the oscillator (resonance radiation) will show no linear polarization.

On the other hand, if the damping is of the same order of

magnitude as the precession velocity, the form of the motion of the oscillator will be given by Fig. 72.

In this case the rosette is incomplete, and shows asymmetry due to the fact that the oscillations have been damped out before a full period of precession takes place. Thus the resulting resonance radiation will be partially polarized (less than in a zero field), and its plane of polarization rotated with respect

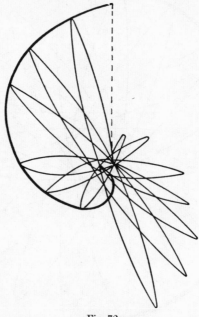

Fig. 72.

to that of the incident light, since the plane of polarization of the resonance radiation will be given by the direction of the maximum electric vector. Making use of the electromagnetic equations of a damped oscillator in a magnetic field and the coherence properties of the light emitted, Breit[2] was able to show that the radiation emitted is partially polarized and its plane of polarization (the plane of maximum light intensity) rotated through an angle ϕ to the X axis. If one measures the polarization by means of an apparatus which keeps the same position with reference to the electric vector of the exciting light throughout the experiment (for example a Wollaston prism), the degree of polarization of the light is given by

$$\frac{P}{P_0} = \frac{1}{1 + \left(\dfrac{eH}{mc} g\tau\right)^2} \qquad \ldots\ldots(179),$$

where P is the polarization observed with a field of intensity H, P_0 that with zero field, τ the mean radiation life of the atom, g a factor to take into account the fact that most atoms do not precess with classical precession velocities but proportional to them, and e, m, c the charge and mass of the electron and velocity of light. On the other hand, one may measure the polarization, as von Keussler did, by rotating a Nicol prism and measuring the maximum and minimum intensities of the light, and use the formula, $P = \dfrac{I_{\max} - I_{\min}}{I_{\max.} + I_{\min.}}$. In this case Breit has shown that the relation between the degree of polarization and the magnetic field is given by

$$\frac{P'}{P_0} = \frac{1}{\sqrt{1 + \left(\dfrac{eH}{mc} g\tau\right)^2}} \qquad \ldots\ldots(180).$$

The rotation of the plane of polarization is given by

$$\tan 2\phi = 2\omega\tau g = \frac{eH}{mc}\tau g \qquad \ldots\ldots(181),$$

where $\omega = \dfrac{eH}{2mc}$, the classical Larmor precession velocity. Thus it is easily seen from Eq. (179) that, as the field increases, the degree of polarization decreases in agreement with Hanle's experiments. Measurements of both effects have been made by Wood and Ellett, Hanle, and von Keussler.

4b. QUANTUM THEORY OF POLARIZATION AND THE ZEEMAN EFFECT. Although the classical theory is able to explain all the polarization phenomena exhibited by the mercury resonance line 2537, it cannot explain the polarization of sodium resonance radiation (as will be shown in a following section). Furthermore, in order to be consistent, a quantum theory explanation must be given.

Hanle[19] was the first to show that the phenomena can be explained on the quantum theory if one considers the Zeeman effect components of the line in question. From the quantum theory of the Zeeman effect it is known that any level of (total angular momentum) quantum number j splits into $2j + 1$ sub-levels in a magnetic field. The sub-levels are designated by a magnetic quantum number m which takes values differing by unity from $-j$ to $+j$. In order for an atom to emit light in jumping from one state to another, the angular momentum quantum number l must change by one unit and m may change by ± 1 or 0. If m does not change ($\Delta m = 0$), the line emitted is analogous to the unshifted or "π component" of the classical Zeeman effect which is polarized parallel to the field. If m changes by one unit ($\Delta m = \pm 1$) the frequency is not the same as that emitted by the atom in a zero field, but differs from it as do the "σ components" of the classical Zeeman effect. These σ components can be shown to be circularly polarized about the field. Although in the weak fields used in experiments on resonance radiation ($0 \to 200$ gauss), the Zeeman components (of Hg for example) are not separated enough to measure except with apparatus of the highest resolving power, the polarization characteristics are clearly defined. This furnishes a powerful means of studying the Zeeman effect of resonance lines whose Zeeman separation is very small.

To explain the experiments on the polarization of the 2537 line of mercury it is necessary to draw a Zeeman diagram of the two states involved. The lower state $6\,^1S_0$ is single with $m = 0$; while the upper state $6\,^3P_1$ is triple with $m = +1$, 0, -1, as shown in Fig. 73. The relative intensities of the several components corresponding to the jump from $6\,^3P_1 \to 6\,^1S_0$ are given at the bottom of the diagram.

Suppose the incident light is polarized parallel to X and contains the frequencies of all the Zeeman components (that is, the exciting line is broadened due to Doppler effect and also due to the magnetic field usually applied to a mercury arc to give an unreversed line). If a strong magnetic field is applied parallel to the X axis and the resonance radiation observed parallel to Y, the electric vector of the incident light is parallel to the field, so that only the π component is absorbed. Since no transitions can occur between the magnetic levels of the $6\,^3P_1$ state, the radiation emitted will be a π component and will have its electric vector parallel to the field. If the field is parallel to the Z axis the σ components will be absorbed in this case and will be consequently re-emitted, and these will be circularly polarized about the field. The observation direction is, however, along the Y axis perpendicular to the field, so that the radiation appears plane polarized parallel to X, as in the corresponding classical case. All other cases discussed under the

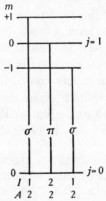

Fig. 73. Zeeman diagram for the line $^1S_0-^3P_1$.

classical theory when the field is perpendicular to the observation direction give analogous results.

Two cases remain to be discussed, (1) when there is no magnetic field and the Zeeman separation vanishes, and (2) when there is a variable field parallel to the direction of observation. In the first case, when there is no magnetic field, all of the different m levels of the upper state fall together and the level is said to be "degenerate". In other words, one cannot tell whether the π or σ components will be excited, since their energy is the same. In order to get round this difficulty Heisenberg[24] suggested the "Principle of Spectroscopic Stability", which postulates that, if a certain degree of polarization is obtained when there is a strong field in the direction of the electric vector of the exciting light, then the same result is obtained on decreasing the field slowly to zero. The reason for choosing the field in the direction of the electric vector is suggested by the classical theory, since the frequency and polarization of a classical oscillator are unchanged when a field is applied parallel to its direction of oscillation. Hanle's and Wood and Ellett's experiments in zero field are thus explained.

Recently Dirac[7] has given a quantum theoretical treatment of an atom in a radiation field which has been very fruitful in giving correct expressions for dispersion and scattering of light. Weisskopf[51] has derived expressions for P and $\tan 2\phi$, on the basis of this theory, for an atom showing normal Zeeman effect, that is, having one single lower magnetic level and three upper ones, as in the case of the 2537 line of mercury. Weisskopf's expressions for this case agree with Eqs. (179) and (181). Recently Breit[4] has shown, by a generalization of Weisskopf's results, that the formulas for the magnetic depolarization of any resonance line are the same as the classical ones if the g factor is taken to be that of the upper level.

5. EXPERIMENTAL VERIFICATION OF THE FORMULAS FOR MAGNETIC DEPOLARIZATION AND THE ANGLE OF MAXIMUM POLARIZATION IN THE CASE OF MERCURY

Using the Nicol prism and photo-cell described above, von Keussler measured the polarization and $\tan 2\phi$ for mercury resonance radiation as a function of the magnetic field applied in the direction of observation. The pressure of the mercury vapour corresponded to $-21°$ C., and the incident radiation was polarized. His results, which will be discussed in detail in § 8, verify in general Eqs. (179) and (181) and lead to a value of the mean life of the $6\,^3P_1$ state of $1\cdot13 \times 10^{-7}$ sec.

Instead of using a magnetic field in the direction of observation, Breit and Ellett[5] and Fermi and Rasetti[14] studied the effect of an alternating field, produced by a vacuum tube and solenoid, on the polarization of mercury resonance radiation. The idea behind the experiment is the following. Suppose the mercury atom to be a damped classical oscillator which will precess about a magnetic field with a Larmor frequency ω. If the reciprocal of the mean life of the oscillator (for Hg, 10^7sec.$^{-1}$) is of the same order of magnitude as the Larmor precession velocity, depolarization will occur in a steady field. If, however, an alternating field of the same strength in gauss is used, and its frequency is much greater than the Larmor frequency, there will be no great effect on the polarization, since the Larmor precession will be first in one direction and then in the opposite direction, depending on the direction of the alternating field, and will be very small in either direction, since the field changes very rapidly. If the alternating frequency is less than the Larmor frequency, the oscillator will have time to precess in the field before the direction of the field changes, and a consequent depolarization will appear.

In Fermi and Rasetti's experiments the magnetic field strength could be varied from 1·13 to 2·13 gauss. (1 gauss gives a Larmor precession velocity of about $1·4 \times 10^6$ sec.$^{-1}$ for a classical oscillator.) The frequency of the field could be changed from 1·2 to 5×10^6 sec.$^{-1}$. At 1·13 gauss and a frequency of 5×10^6 sec.$^{-1}$ they found practically no depolarization, whereas, at 1·87 gauss and the same frequency, depolarization was noted. At 2·13 gauss the depolarization was as large as in a stationary field. If the field strengths had been 3/2 as large as those given, the results could have been explained satisfactorily on the basis of the classical oscillator. The factor 3/2 is just the factor g which gives the relation of the classical Zeeman splitting to that observed; in other words, the precession velocity of the orbital electron of the $6\,^3P_1$ state of mercury is 3/2 the classical Larmor precession velocity. The experiment gives an independent check on the factor g as well as on τ, the mean life in the $6\,^3P_1$ state. The equation for the polarization, as a function of the magnetic field strength and the frequency of the oscillator ν, has been given by Breit[3] and is

$$P = J_0^{\,2}\left(\frac{2p_0}{p}\right) + \frac{2J_1^{\,2}\left(\dfrac{2p_0}{p}\right)}{1+p^2\tau^2} + \cdots,$$

where P is the polarization, $p = 2\pi\nu$, $p_0 = \dfrac{eH}{2mc}g$ and J_n is the Bessel function of order n.

6. POLARIZATION OF SODIUM RESONANCE RADIATION: BREAKDOWN OF CLASSICAL THEORY

6a. EXPERIMENTAL RESULTS ON THE POLARIZATION OF SODIUM RESONANCE RADIATION. The fact that the D line fluorescence of sodium is polarized both in the absence of a magnetic field and with certain orientations of the field, was first shown by Wood and Ellett[54], and the experiments were repeated by Ellett[9] in the hope of getting more quantitative results. In these experiments both the D lines excited the fluorescence, and in observing the emergent light from the resonance tube no attempt was made to separate the two lines.

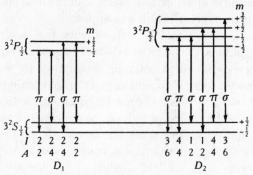

Fig. 74. Zeeman diagram for D lines.

The values they obtained for the polarization, then, are to be taken for the two lines together. Pringsheim and Gaviola[38] used a spectrograph with Savart plate and Nicol prism in observing the polarization of the sodium fluorescence, and found that the D_1 line was always unpolarized and was unaffected by a magnetic field; whereas the D_2 line showed polarization and was affected by a magnetic field. In their experiments the D_2 line never showed more than 25 per cent. polarization. In any theory derived from a classical isotropic oscillator one would expect 100 per cent. polarization for both lines, which is definitely not in agreement with the facts.

6b. THE ZEEMAN LEVELS FOR SODIUM; VAN VLECK'S FORMULAS FOR POLARIZATION. The Zeeman levels for the sodium D lines are given in Fig. 74, and under each line is given the relative intensity I of the line appearing in the pattern, together with the transition probabilities A.

From the considerations of § 4b, it will be seen at once that the D_1 line must be unpolarized, since both upper magnetic levels are connected with both lower levels. The absorption of π components, for example, populates both upper levels equally, and the return from these upper levels to the lower ones entails an emission of π and σ components of equal intensity, so that the resulting radiation is unpolarized. Due to the connection between both upper levels and both lower levels it is clear that the presence of a magnetic field in any direction will leave the polarization unchanged. When the D_2 line, however, is excited with light whose electric vector is parallel to the field (along OX of Fig. 70), the two π components are excited, populating the two middle upper levels. The atom returns to its normal state with the emission of two σ components of intensity 1 each and two π components of intensity 4 each. The polarization observed (along OY) is then

$$P = \frac{8-2}{8+2} = 60 \text{ per cent.}$$

Van Vleck [48] has given formulas for calculating the polarization of any resonance line when the Zeeman *transition probabilities* are known. Consider plane polarized light exciting resonance radiation in a magnetic field H. Let θ be the angle between E and H. Let $A_\pi{}^\mu$ be the transition probability of the linearly polarized Zeeman component originating in the upper magnetic state μ, and $A_\sigma{}^\mu$ be the sum of the corresponding probabilities for the two circularly polarized components. The number of electrons reaching the excited state μ is proportional to $I(A_\pi{}^\mu \cos^2\theta + 1/2 A_\sigma{}^\mu \sin^2\theta)$, where I is the intensity of the incident light. This is true, since absorption and emission probabilities are proportional to each other in non-degenerate systems and since linearly polarized light is only half as effective in exciting circularly polarized as plane polarized components. In returning from the state μ, a fraction of the electrons $\dfrac{A_\pi{}^\mu}{A_\pi{}^\mu + A_\sigma{}^\mu}$ return by linearly polarized transitions, and a fraction $\dfrac{A_\sigma{}^\mu}{A_\pi{}^\mu + A_\sigma{}^\mu}$ by circularly polarized ones. If the resonance radiation is observed at right angles to the field, H, only half the circularly polarized light can be seen. The intensity of the components polarized along and perpendicular to the field is then

$$
\left.
\begin{aligned}
\xi &= IC \sum_\mu \frac{A_\pi{}^\mu}{A_\pi{}^\mu + A_\sigma{}^\mu} (A_\pi{}^\mu \cos^2\theta + \tfrac{1}{2}A_\sigma{}^\mu \sin^2\theta) \\
\eta &= \tfrac{1}{2} IC \sum_\mu \frac{A_\sigma{}^\mu}{A_\pi{}^\mu + A_\sigma{}^\mu} (A_\pi{}^\mu \cos^2\theta + \tfrac{1}{2}A_\sigma{}^\mu \sin^2\theta)
\end{aligned}
\right\}
$$
$$\dots\dots(182),$$

where the sum is to be taken over all Zeeman transition probabilities of the line in question. The polarization is, then,

$$P = \frac{\xi - \eta}{\xi + \eta} \qquad \dots\dots(183).$$

The transition probabilities $A_\pi{}^\mu$ and $A_\sigma{}^\mu$ can be obtained from the Zeeman pattern intensities with the convention that $A_\pi{}^\mu$ = intensity of a π component and $A_\sigma{}^\mu$ = twice the intensity of a σ. The intensity of the π and σ components can in turn be calculated from the sum rule of Ornstein and Burgers.

Since the total chance of leaving any upper magnetic level μ is the same for each level, we may write

$$A_\pi{}^\mu + A_\sigma{}^\mu = A \qquad \dots\dots(184).$$

It is therefore seen at once that the denominators of the expressions for ξ and η in Eq. (182) drop out of the calculation of P in Eq. (183). The expression for the polarization, as given by Van Vleck, may be still further simplified when one remembers that there are $2j + 1$ upper magnetic levels, so that

$$\sum_\mu (A_\pi{}^\mu + A_\sigma{}^\mu) = (2j+1)A \qquad \dots\dots(185).$$

Furthermore, since, for an atom excited by isotropic radiation, the intensity of all the π components must be equal to that of all the σ components, we have

$$\sum_\mu A_\pi{}^\mu - \tfrac{1}{2} \sum_\mu A_\sigma{}^\mu = 0 \qquad \dots\dots(186).$$

Substitution in (183) gives the relation

$$P = \frac{\left(3\Sigma_{\mu} (A_{\pi}{}^{\mu})^2 - \frac{2j+1}{3} A^2\right)(3\cos^2\theta - 1)}{(3\cos^2\theta - 1)\Sigma_{\mu} (A_{\pi}{}^{\mu})^2 + \frac{2j+1}{3} A^2 (3 - \cos^2\theta)}$$

......(187).

This formula has two very interesting consequences. In the first place, if the angle between the electric vector of the exciting light and the applied magnetic field is such that $3\cos^2\theta = 1$ ($\theta = 54°\,45'$) the line will be unpolarized. This will be true for *all resonance lines*, as was first pointed out by Van Vleck. The so-called angle of no polarization has been measured by Hanle for mercury, and by Wood and Ellett for sodium, and found to be in agreement with theory. The second result to be obtained from Eq. (187) is a relation between the polarization to be expected when the electric vector of the exciting light is parallel to the field ($\theta = 0$) and when it is perpendicular to the field ($\theta = \pi/2$). Calling the two polarizations $P_{\parallel}$ and $P_{\perp}$, respectively, one finds the general relation

$$P_{\perp} = \frac{P_{\parallel}}{P_{\parallel} - 2}$$

......(188),

which is again true for all resonance lines.

When resonance radiation consists of two related resonance lines, as in the case of the D lines of sodium, the total polarization of the two lines taken together may be calculated by choosing the transition probabilities in such a way that the chance of leaving any upper magnetic level μ of any upper state j is the same for any such level, and by taking care to bring in the correct relative intensities I of the two lines in the exciting source.

In the case of the D lines of sodium a short calculation shows that the polarization is given by

$$P = \frac{9\cos^2\theta - 3}{7 + 4q + 3\cos^2\theta}$$

......(189),

where q is the ratio of the intensity of D_1 to D_2 in the source. Several interesting cases arise for computation. Let the incident beam travel along OZ (Fig. 70) and the resonance radiation be observed along OY.

(1) It follows from Eq. (187) that the D_1 line alone is unpolarized. This also follows from Eq. (189) by putting $q = \infty$.

(2) For the D_2 line alone, $q = 0$. Let the electric vector of the exciting light and the magnetic field be along X. Then $\theta = 0$ and we have $P = 60$ per cent.

(3) For D_2 alone, let H be parallel to the exciting beam, then $\theta = \pi/2$ and $P = 43$ per cent. This result holds also if the light is unpolarized.

(4) For both lines together, when $q = 1/2$; under conditions of (2) above, $P = 50$ per cent.; and for (3) above, $P = 33$ per cent.

When there is no magnetic field the polarization can be calculated by putting $\theta = 0$, on account of spectroscopic stability. Other interesting cases can be calculated at the pleasure of the reader.

6c. FURTHER COMPARISON OF EXPERIMENT WITH THEORY. In two experiments the polarization of the two D lines was measured separately. Pringsheim and Gaviola[38] separated the two D lines in the fluorescence by means of a spectrograph, while Datta[6] separated the two lines in the exciting beam by means of rotatory dispersion of a quartz crystal and measured the fluorescence visually. Both observers agree that the D_1 line is always unpolarized in accordance with theory.

For the D_2 line alone, Datta made observations of the polarization as a function of the pressure of sodium vapour in the resonance tube. The results of his experiment, made in the absence of a magnetic field, are given in Table XLVI below.

TABLE XLVI

EFFECT OF VAPOUR PRESSURE ON POLARIZATION

Temperatures °C.	Polarization (per cent.)	Pressure of Na (mm. Hg)
150	24	$4 \cdot 8 \times 10^{-6}$
140	26	$2 \cdot 2 \times 10^{-6}$
135	27·5	$1 \cdot 6 \times 10^{-6}$
125	31	8×10^{-7}
115	33	3×10^{-7}

It will be seen from the table that the polarization depends markedly on the vapour pressure of sodium. The explanation of this fact will be fully discussed in a following section, but it is sufficient to point out here that this phenomenon is due to disturbing effects of neighbouring atoms. At the lowest pressures at which Datta could work, the polarization was never more than 33 per cent., whereas the theory predicts 60 per cent. He believed that the polarization-pressure curve was approaching 60 per cent. asymptotically, and that if experiments could be made at lower pressures the theoretical value of 60 per cent. would be obtained. This argument is not convincing when one looks at the polarization-pressure curve in question. Ellett [10], on the other hand, measured the polarization of the two D lines together and found a polarization of 16·3 per cent. This degree of polarization remained the same for vapour pressures of sodium from 3×10^{-7} mm. (115° C.) to $1·9 \times 10^{-8}$ mm. (80° C.), i.e. much lower pressures than used by Datta. Assuming that the exciting source gave the D lines with the ratio $q = 1/2$, Ellett calculated the polarization for D_2 alone to be 20·1 per cent. instead of the 33 per cent. found by Datta. More recent experiments of Heydenburg, Larrick and Ellett [26] give $16·48 \pm 0·33$ per cent. for the two D lines together, and about 20·5 per cent. for the D_2 line alone. Ellett's work shows that decreasing the vapour pressure certainly did not lead to the theoretically expected value of the polarization (for both lines $q = 1/2$, $P = 50$ per cent.).

In this connection, we shall discuss an experiment by Hanle [22], who excited polarized sodium resonance radiation by means of the circularly polarized light of both D lines. The exciting light contained only the left-handed circularly polarized components of the two D lines. The resonance tube was placed in a magnetic field parallel to the exciting light beam and observations were made in a direction nearly parallel to the exciting beam. With this arrangement one would expect only the $-1/2$ and $-3/2$ levels of D_2 and the $-1/2$ level of D_1 to be excited. In fluorescence one should obtain two left-handed circularly polarized components and a linearly polarized component (parallel to the field) from D_2, and one left-handed circularly polarized and one linearly polarized component from D_1. Since the linear components are polarized parallel to the magnetic field and hence to the direction of observation, one should see no light from these, and the resonance radiation should be 100 per cent. polarized (left circular). After making suitable corrections for the fact that observations were made at a small angle to the fluorescent beam, Hanle was able to show that the polarization, in zero field and at the lowest vapour pressures at which the experiment could be carried out (100° C.), was only 60 per cent. instead of 100 per cent. as expected. With large magnetic fields the polarization was not more than 85 per cent.

With a magnetic field parallel to the electric vector of the exciting light Ellett[9] observed 45 per cent. polarization ($H = 60$ gauss) for the two D lines, whereas Datta observed 56 per cent. polarization for the D_2 line alone ($H = 250$ gauss). For large magnetic fields, then, it would appear that the theoretical values of the polarization as given by the Zeeman effect theory have been approached (see § 8f).

We must remark at this point that the polarization found by experiment is usually less than that predicted by the simple Zeeman effect theory. This discrepancy may be due to difficulties in carrying out the experiments or to an over-simplified theory. The experimental difficulties, such as the pressure effect, will be discussed in following paragraphs, as will also theoretical difficulties, such as the effect of hyperfine structure. The discussion in the preceding sections is meant to show the development of the theory in its broad outlines and to correlate polarization measurements with the simple Zeeman effect theory.

7. POLARIZATION OF RESONANCE LINES OF OTHER ELEMENTS: MEAN LIVES OF SEVERAL EXCITED STATES

7a. RESONANCE LINES. The only other elements, with lines having simple Zeeman patterns like mercury, that have been investigated are cadmium and zinc. MacNair[30] and Soleillet[44] investigated the resonance line 3261 ($5\,^1S_0$–$5\,^3P_1$) of cadmium. The former, using rather high vapour pressures (temperature of cadmium reservoir 210° C.) and a sealed-off resonance vessel, found only 30 per cent. polarization. Later, Soleillet found higher values for the polarization. At 210° C. he verified MacNair's value of 30 per cent., but on decreasing the pressure of cadmium vapour to a vapour pressure corresponding to 170° C., he found the polarization to be 73 per cent. and to remain constant as the temperature was lowered to 115° C., in a zero magnetic field. Later experiments, in which the resonance tube was always attached to the pumps, gave 85 per cent. polarization in a zero field or in a field in the direction of the exciting beam (no polarizer was used in these experiments). These last experiments of Soleillet are probably better than the earlier ones, since the resonance tube was freer from any gaseous impurities than in the former. Ellett and Larrick[12], by very careful experiments, found 85–87 per cent. polarization when the vapour pressure of cadmium in the resonance tube corresponded to 146° C. They found that at 168° C. the depolarizing effect first set in. The experiments were performed in a strong magnetic field in a direction parallel to the exciting light beam, which was unpolarized. On the basis of the simple theory the cadmium 3261 line should show 100 per cent. polarization in a zero field or one parallel to the exciting light beam, since the line is entirely analogous to the mercury 2537 line.

TABLE XLVII

CADMIUM 3261, MAGNETIC DEPOLARIZATION

Magnetic field (gauss)	Polarization (per cent.)
0·00	85
0·014	55
0·028	27
0·056	7

Soleillet, using unpolarized exciting light, found that very small fields in the direction of observation of the resonance radiation had a large depolarizing effect. From the data given in Table XLVII, and using $g = 3/2$, Soleillet found the mean life of the $5\,^3P_1$ state to be 2×10^{-6} sec.

Soleillet also investigated the singlet resonance line of cadmium at 2288 and found a polarization of 60 per cent. in the absence of a field. According to the theory, this should also show 100 per cent. polarization. Ellett and Larrick found 76·3 per cent. polarization for this line at low pressures and in a magnetic field parallel to the exciting light beam. Soleillet found, however, that much larger fields in the direction of polarization were necessary to depolarize the radiation than in the case of the 3261 line, as is shown by Table XLVIII.

Table XLVIII

Cadmium 2288, magnetic depolarization

Magnetic field (gauss)	Polarization (per cent.)
0	60
0·03→1	49
25	43
50	38
100	31

From these data he computed a short mean life, of the order of 10^{-9} sec., for the $5\,^1P_1$ state of cadmium. Soleillet[45] also measured the polarization of the two resonance lines in zinc, 3076 and 2139. For the 3076 line, he found a maximum polarization of 67 per cent. in the absence of a field or in a field parallel to the exciting beam. Extremely small fields in the direction of observation were found to depolarize the resonance radiation completely, so that he calculated a mean life of $\tau = 10^{-5}$ sec. for the $4\,^3P_1$ state. The singlet line 2139 showed about 50 per cent. polarization in a magnetic field parallel to the exciting beam.

7b. Line Fluorescence. According to the theory of the classical isotropic oscillator, one would expect that, if resonance radiation were polarized, the direction of polarization would be parallel to the exciting light beam. In direct contradiction to this, and in qualitative agreement with the simple Zeeman theory, are the experiments of Gülke[18] on thallium line fluorescence. According to our discussion of this phenomenon given in Chap. I, we note that excitation is due to the absorption of the lines 3776 and 2768, and that the emitted line fluorescence contains the four lines 3776 ($6\,^2P_{1/2}$–$7\,^2S_{1/2}$), 5350 ($6\,^2P_{3/2}$–$7\,^2S_{1/2}$), 3530 ($6\,^2P_{3/2}$–$6\,^2D_{3/2}$) and 2768 ($6\,^2P_{1/2}$–$6\,^2D_{3/2}$). The Zeeman levels of the states in question are given in Fig. 75 with only one-half the number of transitions drawn. It will be seen at once that, since $6\,^2P_{1/2}$ and $7\,^2S_{1/2}$ have only two magnetic levels each, both magnetic levels of $7\,^2S_{1/2}$ will be equally populated, independent of the polarization of the incident beam. Consequently, the lines 3776 and 5350 emanating from the $7\,^2S_{1/2}$ state will show zero polarization in

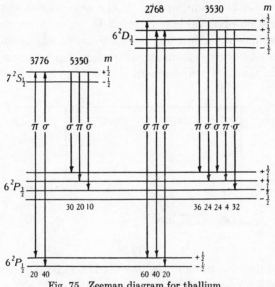

Fig. 75. Zeeman diagram for thallium.

all cases, analogous to the D_1 line of sodium. The $6\,^2D_{3/2}$ state, on the other hand, has four magnetic levels, so that differential population of the upper levels may occur. Making use of the transition probabilities of the various Zeeman components as given in Fig. 75, it is easy to show that, in a zero magnetic field, 2768 should be 60 per cent. polarized parallel to the electric vector of the exciting beam (since the intensity of the π components is greater than of the σ's), and 3530 should be 75 per cent. polarized perpendicular to the direction of polarization of the incident 2768 beam.

Using a polarized light source, and observing the resonance

radiation with a double-image prism and spectrograph, Gülke found that 5350 and 3776 were unpolarized and that 2768 was polarized to the extent of 55 per cent. parallel to the polarization of the exciting beam, and that 3530 was 60 per cent. polarized perpendicular to it, in qualitative agreement with the theory. Gülke also found that high vapour pressures of thallium caused the percentage polarization of both lines to decrease. The figures given above refer to the percentage polarization at the lowest vapour pressure at which the experiment could be made, and probably represent the maximum polarization observable in a zero field.

Formulas for calculating the polarization of any fluorescent line may be derived from considerations similar to those given in § 6 b. Consider a fluorescent line bc (Fig. 76) which may be

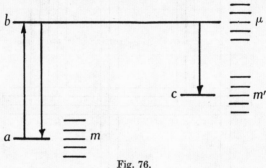

Fig. 76.

excited by absorption of the line ab. Let a given magnetic sub-level of b be designated by μ, and similar magnetic sub-levels of a and c by m and m'. Let the chance of reaching a given magnetic level of b by the absorption of π or σ components of ab be $A_\pi^{ab\mu}$ and $A_\sigma^{ab\mu}$ respectively. Similarly, let the chance of leaving the level μ by a π or σ component of the line bc be $A_\pi^{bc\mu}$ and $A_\sigma^{bc\mu}$. The ξ and η components of the intensity of the line bc will be given by

$$\begin{aligned}
\xi &= I_{ab} C \sum_\mu \frac{A_\pi^{bc\mu}}{A_\pi^{bc\mu} + A_\sigma^{bc\mu}} (A_\pi^{ab\mu} \cos^2\theta + \tfrac{1}{2} A_\sigma^{ab\mu} \sin^2\theta) \\
\eta &= \tfrac{1}{2} I_{ab} C \sum_\mu \frac{A_\sigma^{bc\mu}}{A_\pi^{bc\mu} + A_\sigma^{bc\mu}} (A_\pi^{ab\mu} \cos^2\theta + \tfrac{1}{2} A_\sigma^{ab\mu} \sin^2\theta)
\end{aligned}$$

$$\dots\dots(190).$$

Relations similar to Eqs. (184), (185), (186) hold for ab and bc, so that a calculation shows that

$$P_{bc} = \frac{\left(3\sum_\mu A_\pi{}^{bc\mu} A_\pi{}^{ab\mu} - \frac{2j_b+1}{3} A^{bc} A^{ab}\right)(3\cos^2\theta - 1)}{\sum_\mu A_\pi{}^{bc\mu} A_\pi{}^{ab\mu}(3\cos^2\theta - 1) + \frac{2j_b+1}{3} A^{ab} A^{bc}(3 - \cos^2\theta)}$$

......(191),

and for the important case of $\theta = 0$,

$$P_{bc}(\theta = 0) = \frac{3\sum_\mu A_\pi{}^{bc\mu} A_\pi{}^{ab\mu} - \frac{2j_b+1}{3} A^{bc} A^{ab}}{\sum_\mu A_\pi{}^{bc\mu} A_\pi{}^{ab\mu} + \frac{2j_b+1}{3} A^{bc} A^{ab}}$$

......(192).

In this formula A^{ab} is the total chance of leaving a level μ of b by the path ab, and A^{bc} that by the path bc. If the polarization of only one fluorescent line is to be calculated, one can make $A^{ab} = A^{bc}$ and weight the transition probabilities of the various Zeeman components in such a way as to give this result. When two or more fluoresent lines are not resolved, a case which rarely arises when hyperfine structure is not considered, the transition probabilities of the Zeeman components must be weighted in such a way as to ensure that the relative intensities of the fluorescent lines will conform to the sum rule for intensities.

8. EFFECT OF HYPERFINE STRUCTURE ON THE POLARIZATION OF RESONANCE RADIATION

8a. DETAILED EXPERIMENTAL INVESTIGATION OF THE POLARIZATION OF MERCURY RESONANCE RADIATION. Before going on to a description of further experiments on the polarization of line fluorescence and to a detailed discussion of the effect of collisions and imprisoned resonance radiation on polarization, it will be well to make a more careful examination of the agreement between the observed degree of polarization of the lines so far discussed and that to be expected from the simple theory. Certain discrepancies between experiment and the simple theory are apparent at once. The experimentally observed degree of polarization of the 2537 resonance line of mercury and that of the 3261 resonance line of cadmium are about 80 per cent. and 86 per cent. respectively, whereas the theory predicts 100 per cent. For the D_2 line of sodium, on the other hand, the theoretically predicted degree of polarization is 60 per cent. and that observed not more than 20·5 per cent. It seems certain that these differences are real and are not due to the effect of imprisoned radiation, depolarization by collision, or finite aperture of the exciting and fluorescent beams. Thus, the degree of polarization has been measured as a function of vapour pressure to such low vapour pressures that, in this region, the polarization was found no longer to depend on vapour pressure. It is of interest, therefore, to find a reason for the difference for the existing discrepancy.

Von Keussler [49] suggested that the discrepancy, in the case of mercury, might be due to the hyperfine structure of its resonance line. MacNair [31] measured the Zeeman effect of these h.f.s. components. In a zero field he found the wavelengths of the components to be $-25\cdot4$, $-10\cdot3$, $0\cdot00$, $11\cdot6$, $21\cdot5$ mÅ. In magnetic fields up to 5800 gauss, the latter four became triplets with 3/2 the normal separation, the parallel (π) components of each "line" maintaining the same relative positions as the field is increased. The h.f.s. line at $-25\cdot6$ is not so simple, however. The perpendicular (σ) components of this line behave as the perpendicular components of a 3/2 normal triplet which starts at $-25\cdot6$, but the parallel (π) component increases in wave-length with increasing field. He suggested that the anomalous behaviour of the $-25\cdot4$ mÅ. component might account for the fact that mercury resonance radiation is only 80 per cent. polarized instead of 100 per cent., as would be expected for a 3/2 normal triplet.

Ellett and MacNair [13], therefore, investigated the h.f.s. and polarization of mercury resonance radiation. Their apparatus was as follows: a resonance bulb, containing mercury vapour, was placed in a large coil capable of producing fields from 0 to 3450 gauss. The bulb was radiated from either side by polarized light from two water-cooled and magnetically deflected mercury arcs. The resonance radiation was observed in a direction perpendicular to the two exciting beams by means of a

Wollaston prism and Lummer-Gehrcke plate. The fringe system, formed in the Lummer-Gehrcke plate, from both images from the Wollaston prism was photographed simultaneously on a photographic plate.

In one experiment the mercury vapour pressure corresponded to $-18°$ C., the electric vector of the exciting light was vertical, and the bulb was in zero field. Hyperfine-structure pictures of the resonance radiation, previously shown to be 80 per cent. polarized with E vertical, showed that the vertical component consisted of the four lines $-10·4, 0·00, 11·5$ and $21·5$ mÅ. and the horizontal component of only one line $-25·4$ (perhaps also $21·5$). With the mercury vapour pressure corresponding to $0°$ C. the polarization was less, and the added intensity in the horizontally polarized beam was distributed over all five h.f.s. components. This shows that the depolarization is due to imprisoned resonance radiation or collision. With a strong magnetic field parallel to the exciting light vector, and at the lower mercury vapour pressure, the results remained the same as in the first experiment for fields up to 3450 gauss. Later experiments by Ellett[11] showed that, when mercury resonance radiation is excited by polarized light containing only the two outer h.f.s. components, the polarization is markedly less than 80 per cent.

8b. THEORY OF THE EFFECT OF HYPERFINE STRUCTURE ON THE POLARIZATION OF RESONANCE RADIATION. The formulas for the polarization of a resonance line which consists of several h.f.s. components may be derived with the help of our discussion of the nature of h.f.s. given in Chap. I and the application of the formulas given in §6b of this chapter. Let us first examine the very simple case of a resonance line of an element consisting of only one isotope of nuclear spin i. Each electronic level j will therefore split into a number of hyperfine levels f, and only those h.f.s. components will appear for which the selection rule is obeyed.

Now, it can be shown that, in the presence of a small magnetic field, every hyperfine level f will split into $2f+1$ magnetic sub-levels, such that the quantum number m of each sub-level has values differing by unity and ranging from $f \geqslant m \geqslant -f$. With these considerations in mind, it is a simple matter to construct the Zeeman diagram of any given h.f.s. component. It is apparent that the diagram for a h.f.s. component, having an upper level f_2 and a lower level f_1, will be entirely analogous to that of a gross line not showing h.f.s., and whose upper state has a quantum number j_2 numerically equal to f_2 and a lower state with a quantum number j_1 numerically equal to f_1. The relative intensities and transition probabilities of any Zeeman component can be calculated in a manner analogous to that for a line not showing h.f.s. by substituting f for j in the intensity formulas.

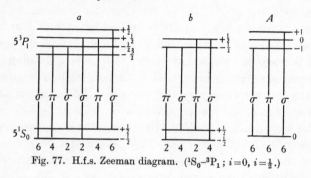

Fig. 77. H.f.s. Zeeman diagram. ($^1S_0$–$^3P_1$; $i=0$, $i=\frac{1}{2}$.)

If a line consist of several h.f.s. components which are not resolved by the apparatus employed, it is necessary to rearrange the transition probabilities of each line so that the relative intensity of each h.f.s. component will be in accord with the sum rule. This may be done in a variety of ways, but is best accomplished by making the total transition probability from any upper magnetic sub-level of any h.f.s. level the same for all such levels. This process is illustrated in Fig. 77, patterns a and b, for a hypothetical line of the form $^1S_0$–$^3P_1$ with $i = 1/2$. The numbers at the bottom of the diagram are the transition probabilities for the Zeeman components in question. It will be noticed that the chance of leaving any upper magnetic sub-level is in this case 6.

We may now calculate the polarization to be expected for the simple case of an element consisting of one isotope having a spin i. Let the lower state have a quantum number j_0 with fine quantum numbers f_a, each divided into magnetic sub-levels designated by m, and the upper state a quantum number j_b with fine quantum numbers f_b, each divided into magnetic sub-levels designated by μ. The number of atoms associated with any hyperfine state f_a of j_0 will be

$$N' = \frac{(2f_a+1)N}{(2j_0+1)(2i+1)},$$

where N is the total number of atoms. The number of atoms in any one magnetic level m of the lower state is

$$N_m = \frac{N}{(2j_0+1)(2i+1)} \qquad \ldots\ldots(193).$$

Now the chance of an atom arriving in an upper state μ of f_b will be proportional to the number of atoms in the lower level N_m, to the intensity of radiation of suitable frequency to excite the h.f.s. component $I_\nu(f_b,f_a)$, and to the transition probabilities $A_\pi^{f_b\mu}$, $A_\sigma^{f_b\mu}$. The chance of reaching a magnetic sub-level μ of f_b is therefore

$$C\frac{NI_\nu(f_b,f_a)}{(2j_0+1)(2i+1)}[A_\pi^{f_b\mu}\cos^2\theta + \tfrac12 A_\sigma^{f_b\mu}\sin^2\theta]$$
$$\ldots\ldots(194).$$

The relative chance of leaving the level μ by a π component is $\dfrac{A_\pi^{f_b\mu}}{A_\pi^{f_b\mu}+A_\sigma^{f_b\mu}}$, with a similar expression for the σ components. The contribution to the intensity of the radiation which is polarized along or perpendicular to the field and comes from the level f_b is

$$\xi_{f_b} = \frac{CNI_\nu(f_b,f_a)}{(2j_0+1)(2i+1)}\cdot\sum_\mu \frac{A_\pi^{f_b\mu}}{A_\pi^{f_b\mu}+A_\sigma^{f_b\mu}}$$
$$[A_\pi^{f_b\mu}\cos^2\theta + \tfrac12 A_\sigma^{f_b\mu}\sin^2\theta]$$
$$\eta_{f_b} = \tfrac12\frac{CNI_\nu(f_b,f_a)}{(2j_0+1)(2i+1)}\cdot\sum_\mu \frac{A_\sigma^{f_b\mu}}{A_\pi^{f_b\mu}+A_\sigma^{f_b\mu}}$$
$$[A_\pi^{f_b\mu}\cos^2\theta + \tfrac12 A_\sigma^{f_b\mu}\sin^2\theta]$$
$$\ldots\ldots(195).$$

The polarization will then be given by

$$P = \frac{\sum\limits_{f_b} (\xi_{f_b} - \eta_{f_b})}{\sum\limits_{f_b} (\xi_{f_b} + \eta_{f_b})} \qquad \ldots\ldots(196).$$

It is more convenient, however, to make use of a quantity which may be called the contribution to the polarization from the level f_b and is defined by

$$P(f_b) = \frac{(\xi_{f_b} - \eta_{f_b})}{\sum\limits_{f_b} (\xi_{f_b} + \eta_{f_b})} \qquad \ldots\ldots(197).$$

It follows from Eqs. (196) and (197) that

$$P = \sum_{f_b} P(f_b) \qquad \ldots\ldots(196a).$$

Using the relations

$$\left.\begin{array}{c} A_\pi^{f_b\mu} + A_\sigma^{f_b\mu} = A^{ba} \\[4pt] \sum\limits_\mu (A_\pi^{f_b\mu} + A_\sigma^{f_b\mu}) = (2f_b + 1) A^{ba} \\[4pt] \sum\limits_\mu A_\pi^{f_b\mu} - \tfrac{1}{2}\sum\limits_\mu A_\sigma^{f_b\mu} = 0 \end{array}\right\} \qquad \ldots\ldots(198),$$

it follows that

$$P(f_b)$$

$$= \frac{I_\nu(f_b, f_a)\, 3\sum\limits_\mu (A_\pi^{f_b\mu})^2 - \dfrac{2f_b + 1}{3}(A^{ba})^2 (3\cos^2\theta - 1)}{\sum\limits_{f_b} I_\nu(f_b, f_a) \sum\limits_\mu (A_\pi^{f_b\mu})^2 (3\cos^2\theta - 1) + \dfrac{2f_b + 1}{3}(A^{ba})^2 (3 - \cos^2\theta)}$$

$$\ldots\ldots(199),$$

and the total polarization P readily follows from Eq. (196a).

In order to calculate the polarization for a line of an element having several isotopes α with relative abundances N_α and nuclear spins i_α, one remembers that the number of isotopes of kind α which are in a given magnetic sub-level m of a lower state is

$$N_m = \frac{N_\alpha}{(2j_0 + 1)(2i_\alpha + 1)}.$$

The polarization formulas then become

$$P(f_b{}^\alpha) = \frac{N_\alpha}{(2i_\alpha + 1)} I_\nu(f_b{}^\alpha, f_a{}^\alpha) \left\{ 3 \sum_\mu (A_\pi{}^{f_b{}^\alpha \mu})^2 - \frac{2f_b{}^\alpha + 1}{3} (A^{ba})^2 \right\}$$
$$\times (3\cos^2\theta - 1)$$
$$\div \sum_\alpha \frac{N_\alpha}{(2i_\alpha + 1)} \cdot \sum_{f_b{}^\alpha} I_\nu(f_b{}^\alpha, f_a{}^\alpha) \left\{ \sum_\mu (A_\pi{}^{f_b{}^\alpha \mu})^2 (3\cos^2\theta - 1) \right.$$
$$\left. + \frac{2f_b{}^\alpha + 1}{3} (A^{ba})^2 (3 - \cos^2\theta) \right\} \quad \ldots\ldots(200)$$

and
$$P = \sum_\alpha \sum_{f_b{}^\alpha} P(f_b{}^\alpha) \qquad \ldots\ldots(201).$$

It follows at once from Eqs. (200) and (201) that the angle of zero polarization is the same for all resonance lines and is independent of h.f.s. The relation, Eq. (188), between $P_\perp$ and $P_\parallel$ is also seen to be general for all resonance lines.

In making a calculation all quantities in Eq. (200) are usually known with the exception of the $I_\nu(f_b{}^\alpha, f_a{}^\alpha)$, which depend on the type of exciting source used. *The polarization of a given resonance line is therefore dependent on the characteristics of the exciting source.* Usually the relative intensity of the h.f.s. components in a given source is not measured in a given experiment. In making the calculation, one of two assumptions is made concerning the relative intensity of the h.f.s. components in the source: (A) All h.f.s. components have the same intensity, $I_\nu(f_b{}^\alpha, f_a{}^\alpha)' = I_\nu(f_b{}^\alpha, f_a{}^\alpha)'' = \ldots$, etc. or (B) the relative intensities of the h.f.s. components depend on the relative abundance of the isotopes of the element in question and the statistical weights of the various h.f.s. states in a calculable way. Vacuum arcs, such as those employed in experiments on resonance radiation, usually show characteristics of the type A. Such excitation is usually called "broad line" excitation. When certain discharge tubes, such as the hollow cathode discharge of Schüler, are used as exciting sources, the intensity characteristics of the exciting source are of the type B, usually designated by "narrow line". Since most of the present experiments on polarization of resonance radiation are performed with vacuum arcs, broad line excitation is usually assumed in making the calculation.

Since a great number of experiments are performed under conditions in which the excitation is of type A, the resonance tube is in zero magnetic field, the incident light is polarized with electric vector vertical, and the observation direction is perpendicular to the exciting beam and the direction of the electric vector, it will be worth while to write down the formula for this case. Here $\theta = 0$, $I_\nu (f_b{}^\alpha, f_a{}^\alpha)' = I_\nu (f_b{}^\alpha, f_a{}^\alpha)''$, etc.:

$$P_0 (f_b{}^\alpha) = \frac{\dfrac{N_\alpha}{(2i_\alpha + 1)} \left\{ 3 \sum\limits_\mu (A_\pi{}^{f_b{}^\alpha \mu})^2 - \dfrac{2f_b{}^\alpha + 1}{3} (A^{ba})^2 \right\}}{\sum\limits_\alpha \sum\limits_{f_i{}^\alpha} \sum\limits_\mu \dfrac{N_\alpha}{(2i_\alpha + 1)} (A_\pi{}^{f_b{}^\alpha \mu})^2 + \sum\limits_\alpha N_\alpha \cdot \dfrac{2j_b + 1}{3} (A^{ba})^2}$$

$$......(202).$$

8c. Theory of the Effect of Hyperfine Structure on the Polarization of Line Fluorescence. The problem of the polarization of line fluorescence may be treated in a similar manner to that of resonance radiation. Consider three atomic energy levels a, b, c in Fig. 76. Let each level be divided into hyperfine levels designated by f_a, f_b, f_c. We wish to calculate the polarization of the line bc excited by the absorption of ab. By the usual calculation it follows that

$$P_{bc} (f_b{}^\alpha)$$
$$= \frac{N_\alpha}{2i_\alpha + 1} \left\{ 3 \sum\limits_\mu A_\pi{}^{f_b{}^{\alpha(a)} \mu} A_\pi{}^{f_b{}^{\alpha(c)} \mu} - \frac{2f_b + 1}{3} A^{ab} A^{bc} \right\} (3 \cos^2 \theta - 1)$$
$$\div \sum\limits_\alpha \frac{N_\alpha}{2i_\alpha + 1} \left\{ \sum\limits_{f_b{}^\alpha} \sum\limits_\mu A_\pi{}^{f_b{}^{\alpha(a)} \mu} A_\pi{}^{f_b{}^{\alpha(c)} \mu} (3 \cos^2 \theta - 1) \right.$$
$$\left. + \frac{(2j_b + 1)(2i_\alpha + 1)}{3} A^{ab} A^{bc} (3 - \cos^2 \theta) \right\} \quad ......(203),$$

if broad line excitation is used. Here $A_\pi{}^{f_b{}^{\alpha(a)} \mu}$ is the chance of reaching the magnetic sub-level μ of $f_b{}^\alpha$ by the absorption of a π component in the line ab; $A_\pi{}^{f_b{}^{\alpha(c)} \mu}$ is the total chance of leaving μ of f_b by making all *possible* π transitions to all states f_c of c. Similarly A^{ab} is the transition probability for any level μ of f_b with regard to the line ab, and A^{bc} is the total chance of leaving any level μ of f_b with regard to the line bc. Since ab is a resonance line, $A_\pi{}^{f_b{}^{\alpha(a)} \mu}$ and A^{ab} may be calculated according to the rules for resonance lines. Since bc is a fluorescent line, care must be taken in calculating $A_\pi{}^{fb^a(c)\mu}$ so that the correct relative intensities of the h.f.s. components of bc will be preserved. In general, several h.f.s. components may be excited by the absorption of one h.f.s. component of ab. Since, however, the gross line ab is usually separated from the gross line bc, it is sufficient to let A^{ab} be numerically equal to A^{bc}.

8d. COMPARISON OF EXPERIMENT WITH PRESENT THEORY OF POLARIZATION. *Cadmium.* Schüler and Keyston[43] have shown that the h.f.s. of the cadmium lines may be accounted for by assuming that the even atomic weight isotopes have no nuclear spin, while those of odd atomic weight have a spin $i = 1/2$. The Zeeman diagrams for the two resonance lines 3261 and 2288 will be similar and are given in Fig. 77. In this diagram the non-spin isotopes are denoted by A and those with spin $i = 1/2$ are denoted by a and b respectively. The adjusted transition probabilities for the π and σ components are given at the foot of each diagram. The calculation of the polarization for various types of exciting sources and of magnetic field has been carried through by Mitchell[33] and more correctly by Ellett and Larrick[12].

Measurements of the polarization of the two lines 2288 and 3261 were made by Ellett and Larrick. The polarization was observed at right angles to the exciting beam, which was unpolarized. The resonance vessel was situated in a magnetic field of 40 gauss parallel to the exciting beam, and the vapour pressure of cadmium in the resonance tube was kept as low as possible (90–105° C. for 2288; 146° C. for 3261). The observed polarization in the case of 2288 was 76·3 per cent. and in the case of 3261 was 86–87 per cent.

The calculation may be carried out with the following assumptions. The absorption coefficient for the line 2288 is high enough, so that the source used may be considered as giving broad lines. In the case of the 3261 line, however, the exciting lamp was operated in such a way (small amounts of cadmium in a hydrogen discharge) that the authors believe that it exhibited narrow line characteristics. If we assume that the h.f.s. component due to the even isotopes (A) coincides with the stronger line (a) of the odd isotopes, then the relative intensity of the exciting lines will be $I_A = I_a = 2 + 3\gamma$; $I_b = 1$, where γ is the ratio of the even to the odd isotopes. One can immediately calculate the polarization to be expected if γ is known. Schüler and Keyston obtained the value $\gamma = 3 \cdot 34$ from their measurements on the h.f.s. of certain cadmium lines. Using this value of γ, $\theta = \pi/2$, and $I_A = I_a = I_b$, Eqs. (200) and (201) give $P = 80 \cdot 5$ per cent. for 2288, which is not in good agreement with the observed value of $76 \cdot 3$ per cent. Taking the observed polarization as correct, a value $\gamma = 2 \cdot 53$ is calculated, which, when used to calculate the polarization of the 3261 line under the above assumptions concerning the source, gives $P = 86 \cdot 1$ per cent., in good agreement with the observed values. The reason for the discrepancy between the value of γ obtained from measurements of the intensities of h.f.s. components of visible cadmium lines and that obtained from the polarization of ultra-violet lines is not apparent.

Mercury. As has been shown in Chap. i, § 8, the h.f.s. of the 2537 resonance line of mercury is complicated by the overlapping of components due to various isotopes, so that the calculation of the polarization becomes rather involved. The calculation has, however, been carried out by Larrick and Heydenburg [29], von Keussler [50] and Mitchell [34]. The Zeeman diagram for all the h.f.s. levels involved is given in Fig. 78. The letters X, A, B, a, etc. above each diagram correspond to the various h.f.s. components given in Fig. 12 of Chap. i. The numbers at the bottom of each Zeeman transition are the transition probabilities for the line. In the case of the isotopes 199 and 201, only half of the transitions are drawn.

In the case of broad line excitation, the intensity of each h.f.s. component is placed equal to unity, and the relative numbers of atoms of given isotopic kinds are given by N_X (even atomic weight) $= 0 \cdot 6988$, $N_{199} = 0 \cdot 1645$, $N_{201} = 0 \cdot 1367$. For narrow line excitation, the relative intensities of the various components are given in Fig. 12 of Chap. i. The calculation is then carried out with the help of Eqs. (200) and (201). The results of the calculation are given in Table XLIX.

The results are in substantial agreement with experiment. Von Keussler, using the usual mercury arc, found 79·5 per cent. polarization, while Olson [37] varied the current in his source and found 79 per cent. with a current of 3·5 amperes, 84 per

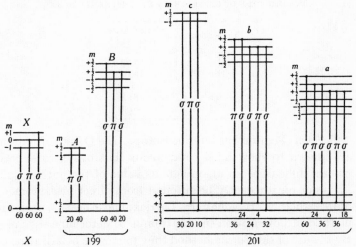

Fig. 78. Zeeman diagram for 2537 (showing h.f.s.).

TABLE XLIX

POLARIZATION OF 2537

Excitation	Polarization (per cent.)	
	$\theta = 0$	$\theta = \pi/2$
Broad line	84·7	73·5
Narrow line	88·7	81·2

cent. at 1 ampere, and 86 per cent. at 0·4 ampere. Both experiments were performed in a zero magnetic field ($\theta = 0$).

The experiments of Ellett and MacNair [13], on the polarization of the separate h.f.s. components of 2537, showed that the three inner components (11·5, 0, − 10·4 mÅ.) were practically completely polarized, whereas the two outer components showed incomplete polarization. Table L gives the results of the calculation for the polarization of each h.f.s. component in zero magnetic field. The agreement between theory and experiment may be considered as satisfactory.

TABLE L

POLARIZATION OF SEPARATE H.F.S. COMPONENTS OF 2537

Component	Polarization (per cent.)
21·5	55·9
11·5	100
0·0	100
− 10·4	84·8
− 25·4	51·4

Sodium. Sodium has but one isotope. The D lines of sodium are known to show h.f.s. components, but owing to experimental difficulties an unambiguous value of the nuclear spin i has not until recently been found from direct measurement of its structure. Heydenburg, Larrick, and Ellett[26] used the method of polarization in an attempt to determine the spin. By the use of the Cornu method they found the polarization of the two D lines together to be $16\cdot48 \pm 0\cdot33$ per cent., and of the D_2 line alone to be $20\cdot5$ per cent. They calculated the polarization to be expected for various values of the nuclear moment, and their results are given in Table LI below. The value $i = 3/2$

TABLE LI

SODIUM D LINES

i	P_{D_2}	$P_{D_2+D_1}$
0	60	50
1/2	40·54	33·33
1	20·50	16·6
3/2	18·61	15·07
2	17·33	14·02
5/2	16·47	13·32
3	15·91	12·84
7/2	15·54	12·56
4	15·28	12·35
9/2	15·09	12·19
5	14·28	11·54
Observed		$16\cdot48 \pm \cdot33$

has been obtained from three independent methods by Rabi and Cohen[39], Granath and Van Atta[17], and Urey and Joffe[47]. It appears certain, from the work of these investigators, that the value $i = 3/2$ is the correct one. It may be seen from Table LI that the experiment appears to be in agreement with a value $i = 1$. Owing to the fact that the separations of the upper h.f.s. levels are small, certain corrections have to be made to the calculation. The necessary corrections have been pointed out by Breit[4], and a calculation, performed by Heydenburg and Ellett, appears to show that the experimentally observed value of the polarization is in accord with a value of $i = 3/2$.

Thallium. Ellett[10] has carried through similar calculations for the slightly more complicated cases of thallium resonance radiation and line fluorescence. He used the h.f.s. data of Schüler and Brück[42] who found a moment $i = 1/2$, and made the calculation for a magnetic field parallel to the electric vector of the exciting light (and hence also for a zero field). He further assumed that the two isotopes (203 and 205) behaved as one entity with a moment $i = 1/2$. In calculating the relative intensities of the h.f.s. components for a "narrow" line source, he used those obtained from the h.f.s. sum rule. The results of the calculation are compared with experiment in the following table.

TABLE LII

Line	Polarization (per cent.)			
	Nuclear moment			Observed
	$i = 0$	$i = 1/2$		
		A	B	
2768	+60	+33·2	+35·1	+55 (470° C.)
3530	−75	−41·8	−48·8	−60 (470° C.)
3776	0	0	0	0
5350	0	0	0	0

+ means parallel to electric vector of exciting light.
− means perpendicular to electric vector of exciting light.
A, broad line excitation; B, narrow line.

The experiments used for comparison in the case of 2768 and 3530 are those of Gülke[18], taken at the lowest pressure at which he worked. The experiments on 3776 and 5350 are due to Ellett [8]. It will be seen from the table that the polarization of 3776 and 5350 is always zero, independent of the assumed nuclear moment, and the results agree with experiment. The quantitative agreement between theory and experiment for the two polarized lines is, however, not good.

8e. EFFECT OF HYPERFINE STRUCTURE ON MAGNETIC DEPOLARIZATION AND THE ANGLE OF MAXIMUM POLARIZATION. It is of interest now to discuss the effect of h.f.s. on the magnetic depolarization of resonance radiation, especially since the problem has an important bearing on the value of the mean life of a given state as measured by the depolarization of a fluorescent line coming from that state. We have seen that the classical theory gave a formula connecting the mean life with the polarization of a line when it was measured in a weak field parallel to the direction of observation. Recently Breit [4] has applied the quantum theory of radiation to the problem and has derived formulas for the magnetic depolarization and angle of maximum polarization of a resonance line in weak magnetic fields.

If plane polarized light is incident on a resonance tube, and the resonance radiation is observed in a direction at right angles to both the exciting beam and its electric vector, and if a weak magnetic field H is applied parallel to the direction of observation, the polarization as a function of the field is given by

$$P(H) = \sum_{\alpha, f_b{}^\alpha} \frac{P_0(f_b{}^\alpha)}{1 + \left(\dfrac{eH}{mc} g_{f_b{}^\alpha} \tau\right)^2} \qquad \ldots\ldots(204).$$

In the formula, $P(H)$ is the polarization of a resonance or fluorescent line in a field H, $P_0(f_b{}^\alpha)$ is the polarization in zero field of the components coming from an upper hyperfine state of an isotope α, $g_{f_b{}^\alpha}$ is the hyperfine g-factor for that state, and τ is its mean life. The values of $P_0(f_b{}^\alpha)$ may be obtained for any line from Eq. (202), or Eq. (203) in the case of fluorescent lines, and the g-values in question may be calculated from the usual formula*. The angle of rotation of the

* See L. Pauling and S. Goudsmit, *The Structure of Line Spectra*, McGraw Hill Book Company, p. 219.

plane of polarization, i.e. the angle of maximum polarization ϕ, is given by

$$\tan 2\phi = \frac{\sum\limits_{\alpha, f_b{}^\alpha} \sin 2\phi\,(f_b{}^\alpha)\cos 2\phi\,(f_b{}^\alpha)}{\sum\limits_{\alpha, f_b{}^\alpha} P_0\,(f_b{}^\alpha)\cos^2 2\phi\,(f_b{}^\alpha)} \quad \ldots\ldots(205),$$

where

$$\tan 2\phi\,(f_b{}^\alpha) = \frac{eH}{mc}\,g_{f_b}{}^a\tau \quad \ldots\ldots(206).$$

Mitchell[35] has used these formulas to calculate the mean lives of several excited states from already existing data for $P(H)$ and $\tan 2\phi$, and to see what error may have been made in τ owing to the neglect of h.f.s. In carrying out the calculation, it is assumed that τ is the same for any h.f.s. state of each isotope and is equal to the mean life of the atom in the excited state (n, l, j). This assumption has theoretical justification, although a direct experimental proof of it is lacking. Indirect evidence as to the validity of the assumption is, however, contained in the fact that calculations made using the assumption are in accord with experimental facts, as will be evident from the following discussion.

The calculation has been made for the 2537 resonance line of mercury, using the data given in Table LIII together with the value of $\tau = 1\cdot08 \times 10^{-7}$ sec. This value of τ is that obtained by Garrett, and has been shown to be in agreement with absorption coefficient data by Zemansky and Zehden (see Chap. III). The values of $P_0\,(f_b{}^\alpha)$ given in the table are calculated with the help of Eq. (202). The results of the numerical calculation of $P(H)$ and $\tan 2\phi$ are plotted in Fig. 79 and Fig. 80. The upper curve of Fig. 79 gives $P(H)$ as a function of magnetic field obtained from Eq. (204). The lower curve is a hypothetical curve obtained on the assumption that the line 2537 was due to isotopes having no nuclear spin and with a g-value for the $6\,^3P_1$ state of $3/2$, but that, for some unknown reason, the polarization in zero field was only $84\cdot7$ per cent. instead of the expected 100 per cent. Some experimentalists have made exactly this assumption in calculating τ for mercury and cadmium resonance radiation. The lower curve,

TABLE LIII

State	$P_0(f_b{}^a)$	$g(f_b{}^a)$
Even $i=0$ $f_b=1$	0·754	3/2
199 $i=1/2$ $f_b=1/2$	0·000	—
199 $i=1/2$ $f_b=3/2$	0·058	1
201 $i=3/2$ $f_b=1/2$	0·000	—
201 $i=3/2$ $f_b=3/2$	0·016	2/5
201 $i=3/2$ $f_b=5/2$	0·020	3/5
	$P_0=0·848$	

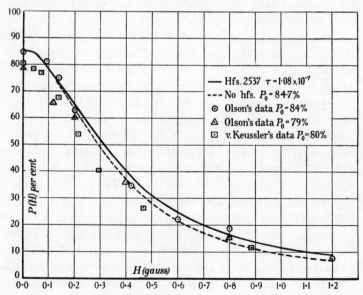

Fig. 79. Polarization of 2537 as a function of weak magnetic fields.

therefore, shows the error in such a procedure for the case of 2537; and it is seen to be quite small. The experimental points plotted for comparison with the theory are those obtained by Olson[37] and by von Keussler[49]. The circles represent experiments made by Olson with his mercury arc operating under such conditions that the polarization in zero field is that given by theory. It will be seen that the points fit the theoretical curve, with $\tau = 1 \cdot 08 \times 10^{-7}$ sec., within the limit of experimental error. Olson obtained a value for τ of $0 \cdot 98 \times 10^{-7}$ sec. from a method of handling his results which did not involve the consideration of h.f.s. The triangles on the diagram represent experimental points in which the exciting arc was run with higher current densities, while the squares give von Keussler's results. It is to be noted that, in this case, in zero field the experimentally observed polarization is not in accord with that predicted by theory. As the depolarizing field is increased, however, the points lie well on the theoretical curve. This is probably due to the self-reversal of the h.f.s. components in the arc operated at high current densities.

In Fig. 80 the lower curve gives $\tan 2\phi$ as a function of the magnetic field H, calculated from Eq. (205). The upper curve is that obtained assuming that the entire radiation of the line 2537 is due to non-spin isotopes having a g-value 3/2. The experimental points are those of von Keussler, from which he calculated the value of $\tau = 1 \cdot 13 \times 10^{-7}$ sec. The disagreement between experiment and theory is probably due to the intensity distribution in the source.

Mrozowski[36] has measured the angle of maximum polarization as a function of magnetic field for certain separate h.f.s. components of the line 2537. In particular he has made measurements on the $0 \cdot 0$ and $+ 11 \cdot 5$ mÅ. components together (each due to non-spin isotopes) and the $- 25 \cdot 4$ mÅ. component. The latter component is due to the lines A and c coming from isotopes 199 and 201, respectively (see Fig. 12). The line c is, however, unpolarized, so that the contribution to the polarization, and hence to the angle of maximum polarization, is due entirely to the component A. The g-value for the upper state corresponding to the $0 \cdot 0$ and $+ 11 \cdot 5$ mÅ. components is 3/2, while that for the component A is 1. In these special cases Eq. (206) reduces to

$$\tan 2\phi = g_{f_b{}^a}\tau \cdot \frac{eH}{mc}.$$

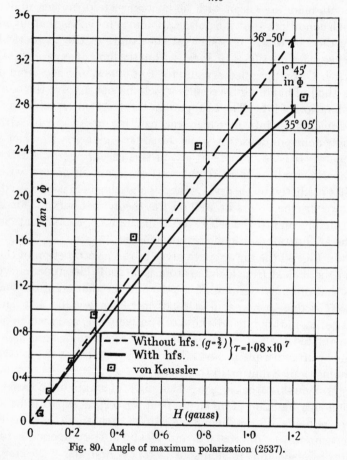

Fig. 80. Angle of maximum polarization (2537).

If one plots, therefore, $\tan 2\phi$ against H, for the components in question a straight line should result. If τ is the same for both isotopes, the slopes of the lines should be in the ratio of the $g(f_b{}^\alpha)$ for the upper states, or in our case as 3/2:1. Plots of Mrozowski's points for $\tan 2\phi$ against H do yield straight lines, and the ratio of the slopes for the central components to that for the $-25\cdot4$ mÅ. component is $1\cdot50$. This value is in exact agreement with theory, and would appear to show that τ is the same for both isotopes and is not dependent on which state f or j is involved as the upper state.

A similar calculation to that given above was made for cadmium by Mitchell. The results of a plot of $P(H)$ against H give results in fair agreement with Soleillet's values, listed in Table XLVII, if τ is taken to be $2\cdot5 \times 10^{-6}$ sec.

8f. EFFECT OF LARGE MAGNETIC FIELDS; PASCHEN-BACK EFFECT OF HYPERFINE STRUCTURE. In our treatment of the effect of h.f.s. and polarization we have supposed, so far, that the vectors **i** and **j** are coupled together to form a resultant **f**, and that, in a magnetic field, this vector may have $2f+1$ projections on the field direction, giving rise to the magnetic quantum number m_f, which we have previously called simply m. If now the magnetic field is sufficiently strong, the coupling between **i** and **j** will be broken, and each vector will orient itself in the field independently of the other. Such an effect is called the complete Paschen-Back effect of h.f.s. We may call these projections m_i and m_j respectively, and their vector sum m. The selection rules which now apply are: (1) m_i does not change; (2) $\Delta m_j = 0, \pm 1$. The polarization rules in this case are now given by the change in m_j ($\Delta m_j = 0$, π components; $\Delta m_j = \pm 1$, σ components). A schematic diagram for a line of the type $^1S_0$–$^3P_1$ with $i = 1/2$ is shown in Fig. 81, which should be compared with Fig. 77, showing the same line in weak fields. It may be seen from Fig. 81 that, if an atom, situated in a very strong magnetic field, absorbs radiation which is polarized parallel to the field, the levels with $m_j = 0$ will be reached. Since no transitions can occur to levels with m_j different from zero, π components will accordingly be re-radiated. One would expect, therefore, that a resonance line showing this type of structure would be 100 per cent. polarized if observed in a very strong magnetic field. From the above discussion it follows that the effect of h.f.s. on the polarization of resonance radiation disappears if the experiment is carried out in very strong fields. This will be true for any resonance or fluorescent line, so that the degree of polarization calculated by means of the simple theory should be in agreement with experiment, provided the magnetic field is large enough.

The strength of field necessary to obtain complete Paschen-Back effect of h.f.s. depends on the ratio of the separation of the h.f.s. levels in zero field to that due to the Zeeman effect on the multiplet level. When this ratio is large compared to unity, the Paschen-Back effect of h.f.s. will be complete. If, on the other hand, the ratio is of the order of unity, intermediate coupling schemes must be used. In this case formulas have

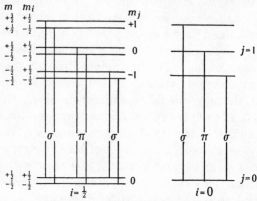

Fig. 81. Paschen-Back effect of h.f.s.

been developed giving the transition probabilities for various π and σ components as a function of field strength*.

This effect has been shown by various experiments. Von Keussler[49] showed that, whereas mercury resonance radiation was about 80 per cent. polarized in fields ranging from 0 to 500 gauss, the polarization increased to about 100 per cent. in a field of 7900 gauss. Systematic determinations of the polarization of sodium resonance radiation as a function of magnetic field strength have been carried out by Larrick[28]. In his experiment, the D lines were excited by polarized light and the resonance tube placed in a strong magnetic field parallel to the direction of the electric vector of the incident

* See L. Pauling and S. Goudsmit, *The Structure of Line Spectra*, McGraw Hill Book Company, p. 219.

light. The polarization of both D lines together was measured by observing in a direction at right angles to the magnetic field and to the direction of the incident light beam. The results of the experiment, showing degree of polarization as a function of field strength, are given in Table LIV. It may be seen from

TABLE LIV

POLARIZATION OF D LINES IN STRONG FIELDS

H	$P_{D_1+D_2}$	Error
0	16·48	±0·38
10	16·26	0·30
20	21·37	0·38
50	34·54	0·33
70	38·86	0·33
90	43	—
170	44·5	—
315	46·25	—

the table that the degree of polarization increased from 16·48 per cent. in zero field to 46·25 per cent. in a field of 315 gauss. Since we have shown that the degree of polarization of the two D lines together should be 50 per cent., if calculated on the basis of no h.f.s., it seems probable, from Larrick's result, that a field of 315 gauss has caused an almost complete Paschen-Back effect of the h.f.s. of the D lines. Larrick used the theoretical formulas for calculating the degree of polarization in the intermediate field range, and, by comparing theory with experimental results, has determined certain constants having to do with the nuclear spin of sodium.

Heydenburg [25] performed a similar experiment on the 2288 line of cadmium. He excited the resonance line with unpolarized light, applied strong magnetic fields in the direction of the incident light beam, and measured the polarization of the resonance radiation (observed in a direction perpendicular to the field) by the method of crossed Wollaston prisms. Table LV, taken from his paper, shows the results of the experiment. In column 1 is given the observed polarization for the lines from both even and odd isotopes, and in column 2 the observed data of column 1 have been corrected for angular aperture of the exciting light beam. By the methods given in this chapter, the degree of polarization for the odd isotopes alone was calculated from that observed for the total radiation. By use of Eq. (188) $P_{\parallel}$ for the odd isotopes was calculated from the data in column 3. The results show that the radiation from the odd isotopes is about 43 per cent. polarized ($P_{\parallel}$) in a zero field and increases to about 90 per cent. in a field of 563 gauss. From the intermediate field data he was able to calculate the constants of Goudsmit's equation.

TABLE LV

POLARIZATION OF Cd 2288 IN STRONG FIELDS

P observed	P_0 corrected	$P_\perp$ odd isotopes	$P_{\parallel}$ odd isotopes	H (gauss)
%	%	%	%	
76·3	76·7	27·3	42·7	0
76·8	77·3	28·8	44·7	75
79·3	79·9	36·0	53·0	144
81·9	82·5	43·4	60·5	200
85·5	86·2	54·4	70·5	255
88·5	89·1	63·4	77·6	315
91·3	92·1	73·0	83·7	375
94·0	95·0	82·6	90·5	563

9. STEPWISE RADIATION

9a. POLARIZATION OF STEPWISE RADIATION. The stepwise radiation of mercury (see Chap. II), notably the visible triplet $6\,^3P_{012}-7\,^3S_1$ and the ultra-violet triplet $6\,^3P_{012}-6\,^3D_1$, has recently been shown to exhibit polarization if excited by polarized light from a mercury arc. The phenomenon of polarization of stepwise radiation was discovered in a rather striking way by Hanle and Richter[23]. They sent light from a water-cooled quartz mercury arc through a calcite block, and focused the two images as beams in the resonance tube containing mercury vapour and about 2 mm. of nitrogen. They noticed that the two beams, seen as fluorescence, were of different colours; the beam with electric vector parallel to OX being blue green, the other (parallel to OY) being yellowish green. They immediately recognized this phenomenon as being due to polarization.

Quantitative investigations of the polarization of the visible lines were made by Hanle and Richter, and of both ultra-violet and visible lines by Mitchell[32] and Richter[41]. The results of Richter's observations are given in Table LVI.

The calculation has been carried through by Mitchell[35]* using Eq. (203) and the h.f.s. analysis of the mercury spectrum of Schüler and Keyston. It will be sufficient to mention certain assumptions of the calculations. The main contribution to the intensity of the stepwise lines is due to the non-spin isotopes, since they constitute 70 per cent. of the mixture. Since the calculation involving h.f.s. is rather tedious, we may discuss

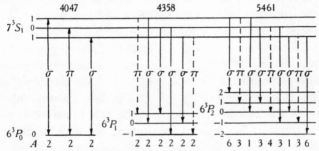

Fig. 82. Zeeman levels of visible triplet of Hg. (Even isotopes.)

the problem first in terms of the non-spin isotopes and then make the necessary corrections to account for the h.f.s.

The magnetic energy level diagrams for the various states of the even isotopes and the allowed Zeeman transitions for the several lines in question are given in Fig. 82. It will be noticed at once that the $6\,^3P_0$ state is single so that, as far as any subsequent absorption process is concerned, the way in which the $6\,^3P_0$ state was reached is of no consequence. The polarization of the original 2537 line, therefore, need not be taken into account, nor need the effect of the collision between the $6\,^3P_1$ mercury atom and the nitrogen molecule introduce any complications. The $6\,^3P_0$ level may then be treated as the ground level and the phenomenon resolves itself into a simple case of

* Earlier calculations by Mitchell [34] and von Keussler [50] led to erroneous numerical values for the polarization due to incorrect methods of calculation. The correct values for the polarization are given in Table LVI.

the polarization of line fluorescence. Since the $7\,^3S_1$ and $6\,^3D_1$ levels have the same type of magnetic splitting, and since the three ^3P states are involved in either the visible or ultra-violet triplet, the polarization of the visible lines ending on the various ^3P states will be the same as the ultra-violet lines ending on the same states.

Starting then with atoms in the $6\,^3P_0$ state, illumination with 4047 (or 2967) (Electric vector parallel to the field, i.e. $E \parallel OX \parallel H$ of Fig. 70) leads to the absorption of the π component of 4047 (or 2967), and to the population of the middle level of the $7\,^3S_1$ ($6\,^3D_1$) state. From this state the π component of 4047 (2967), two σ components of 4358 (3131), and various π and σ components of 5461 (3663), may be re-radiated. A short calculation shows that, under these circumstances, 4047 (2967) should be $+100$ per cent., 4358 (3131) -100 per cent., and 5461 (3663) $+14\cdot3$ per cent. polarized, when observations are made perpendicular to the field and to the direction of the incident radiation (along OY of Fig. 70). These results also hold in the absence of a field. If the field is perpendicular to both the electric vector of the exciting beam and to the observation direction ($\parallel OZ$), then one would expect 100 per cent., -33 per cent., 7 per cent. for the three lines 4047, 4358 and 5461 respectively. A sufficiently strong field parallel to the observation direction should give zero polarization for all lines. In the above notation the $(+)$ means polarization parallel to that of the exciting beam and $(-)$ perpendicular to it.

In considering the effect of h.f.s. one sees immediately that there is more than one magnetic level connected with the $6\,^3P_0$ state. The effect of the nitrogen on the transfer of atoms from the $6\,^3P_1$ to the $6\,^3P_0$ state must therefore be re-investigated. If one assumes an equal mean radiation life for each h.f.s. state of the resonance line (2537), and that the chance of transfer to the $6\,^3P_0$ state by collision with nitrogen is the same for all isotopes, it follows that the relative number of isotopes in the $6\,^3P_0$ state after the collision process will be the same as that in the ground state ($6\,^1S_0$). It must be assumed further that collision with nitrogen leads to an equal distribution of atoms among the magnetic sub-levels of a given hyperfine state.

Since $6\,^3P_0$ has the same structure as the ground level, the distribution of atoms among the magnetic sub-levels may be taken to be the same as in the ground state. With these assumptions in mind, the Zeeman diagrams of the h.f.s. components of the various lines in question may be drawn and the calculation made with the help of Eq. (203). Table LVI shows the result of the calculation, together with Richter's experimental results, taken in zero field.

TABLE LVI

| Line | Polarization | | ϕ (Obs.) | τ (Richter) | τ (Calc.) |
	Obs.	Calc.			
4047	72 ± 6	$84 \cdot 7$	$17°$	$4 \cdot 8 \times 10^{-8}$	$7 \cdot 2 \ \times 10^{-9}$
4358	-49 ± 6	$-67 \cdot 0$	$29°$	$4 \cdot 6 \times 10^{-8}$	$1 \cdot 69 \times 10^{-8}$
5461	13 ± 1	$8 \cdot 6$	$29 \cdot 5°$	$1 \cdot 7 \times 10^{-7}$	$1 \cdot 53 \times 10^{-8}$
2967	67 ± 7	$84 \cdot 7$	—	—	—
3131	-29 ± 7	$-67 \cdot 0$	—	—	—
3663	42 ± 4	$8 \cdot 6$	—	—	—

$H = 2 \cdot 81$ gauss. The minus sign (–) indicates that the line is polarized with its electric vector at right angles to the electric vector of the incident light.

One may see from the table that the agreement between theory and experiment is not good. The phenomenon is a complicated one, and further work must be done to clear up present existing difficulties. Such work is now in progress, and it is to be hoped that it will lead to a successful solution of the various difficulties.

9*b*. MEAN LIFE OF THE $7\,^3S_1$ STATE OF Hg. By applying small magnetic fields in the direction of observation and keeping the nitrogen pressure constant, Richter measured the magnetic depolarization and rotation of the plane of polarization of the visible triplet. From the angle of rotation, ϕ, Richter calculated the mean life of the $7\,^3S_1$ state as measured by 4047, 4358, 5461 to be $\tau_{4047} = 4 \cdot 8 \times 10^{-8}$ sec., $\tau_{4358} = 4 \cdot 6 \times 10^{-8}$ sec., $\tau_{5461} = 1 \cdot 7 \times 10^{-7}$ sec.

These figures are only qualitative, since the mean life was found to depend on the nitrogen pressure. The interesting thing, however, is that the mean life of the $7\,^3S_1$ state as measured by 4047 and 4358 at $1 \cdot 77$ mm. nitrogen pressure is the same, but as measured by 5461 is four times larger.

These results, if correct, would be in disagreement with present theories of atomic structure. One would expect, theoretically, the mean life of a given electronic state n to be a property of that state, since by definition $\tau_n = (\sum_k A_{nk})^{-1}$. If the mean life of a state n be measured by experiments, such as those described above, on individual lines coming from that state the result should be the same for all such lines.

In making the calculation from the measured angular rotation ϕ, Richter neglected the effect of hyperfine structure and also used incorrect formulas for the calculation, and furthermore made a numerical mistake in the computation. Mitchell has recalculated Richter's results with the help of Eq. (205) and the hyperfine structure data and has obtained results quite different from those given above. These are shown in the last column of Table LVI. It appears from the results that further measurements will have to be made in order to be sure of the mean life of this state. One must point out that the apparent confirmation of Randall's[39a] results by Richter must be considered to be no confirmation at all, due to an erroneous method of handling the experimental data.

10. DEPOLARIZATION BY COLLISION

We have seen in the foregoing sections that the polarization of resonance radiation decreases when the vapour pressure increases, and have ascribed this phenomenon to the disturbing effect of neighbouring particles on the emitting atom. Leaving aside the question of the depolarizing effect of high vapour pressure of an element on its own resonance radiation—a problem which offers some difficulties—it will be well to consider first the effect of foreign gases on the polarization of mercury and sodium resonance radiation.

In the first place, Wood measured the polarization of mercury resonance radiation as a function of the pressure of added gas, using the character of the polarization fringes from a Savart plate arrangement as a measure of the polarization. He found that with 0·65 mm. of air in the resonance tube the polarization fringes were strong and with 1 cm. of air they disappeared entirely; with 2 mm. of helium they were faint, while with 1 cm. of hydrogen they were strong. These results were only qualitative and showed the influence of added gases on the polarization.

Quantitative experiments were performed by von Keussler[49]. With the mercury vapour pressure at 2×10^{-5} mm. ($-21°$ C.) various pressures of several foreign gases (O_2, H_2, A, He, N_2, CO_2, H_2O) were added and the percentage polarization measured as a function of the gas pressure. The results of the observations are given in Fig. 83, from which it will be

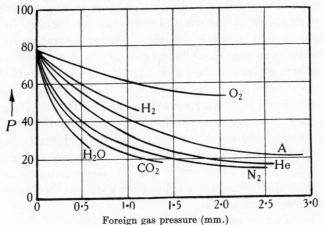

Fig. 83. Depolarization by collision.

noticed at once that O_2 and H_2 have the smallest depolarizing effect, while H_2O and CO_2 have the largest, with the other gases ranging in between in the order named above.

In Chap. IV it was shown that H_2 and O_2 are the most efficient in quenching mercury resonance radiation. Other gases are much less efficient in removing mercury atoms from the $6\,^3P_1$ state, and kinetic theory calculations showed that, in some cases, a number of collisions were necessary before the $6\,^3P_1$ state became depopulated. Qualitatively, the low depolarization action of H_2 and O_2 results from the fact that when an excited mercury atom is struck by an H_2 or O_2 molecule it practically always loses its power to radiate, and the remaining unstruck atoms, being practically undisturbed, can radiate light showing a high degree of polarization. With argon, on the other hand, the excited mercury atom may endure several collisions without losing its power to radiate, but when it does emit radiation the polarization properties will have been destroyed by these collisions.

The mechanism of depolarization is usually thought of in terms of the Zeeman levels. In small fields the Zeeman levels are not widely separated and the energy difference between them is small. A colliding atom may give or lose a certain amount of kinetic energy to the excited atom, transferring it from one Zeeman level to another and thus causing a decrease of polarization of the emitted radiation. An increase in the magnetic field, however, causes a greater separation of the Zeeman levels, and the colliding atom is then not so efficient in transferring excited atoms from one magnetic state to another, due to the greater energy required, so that the polarization is not decreased to such a great extent by the addition of a given pressure of foreign gas. This has been shown by Hanle [22] for the case of sodium resonance radiation and argon, neon, helium and hydrogen. In every case, it took a greater pressure of added gas to depolarize the resonance radiation a given amount when there was a field of 600 gauss on the resonance tube, than in zero field.

The effect of collision on the depolarization of resonance radiation may be represented by a formula similar to the Stern-Volmer formula, namely

$$P = \frac{P_0}{1 + \tau Z_D} \qquad \dots\dots(207),$$

where P is the polarization observed with foreign gas pressure p, P_0 the polarization when no foreign gas is present, and τZ_D is the number of depolarizing collisions per lifetime of the excited atom. Since in von Keussler's experiments the mercury vapour pressure was kept very low (corresponding to a temperature $t = -21^\circ$ C.), the effect of imprisoned resonance radiation may be neglected. It has been shown in §1 of Chap. IV that

$$Z_D = 2N\sigma_D{}^2 \sqrt{2\pi RT \left(\frac{1}{M_1} + \frac{1}{M_2}\right)} \qquad \dots\dots(208),$$

where $\sigma_D{}^2$ can be regarded as an effective cross-section for depolarization. From von Keussler's experimental results and the use of Eqs. (207) and (208) values of $\sigma_D{}^2$ have been found and are given in Table LVII. By comparing these figures with

TABLE LVII

DEPOLARIZING CROSS-SECTION FOR Hg RESONANCE RADIATION

Foreign gas	$\sigma_D{}^2 \times 10^{16}$
O_2	5·28
H_2	3·29
CO_2	49·5
H_2O	47·7
N_2	33·8
He	9·92
A	17·6

$\sigma_Q{}^2$ (Chap. IV, § 8), it will be seen that there is no relation between the effective cross-section for depolarization and that for quenching. It may be seen from Table LVII that the values of σ_D are of about the same order of magnitude as the usual kinetic theory diameters. Abnormally large values of σ_D (10^4 times the kinetic theory value) were found by Datta for the depolarization of sodium resonance radiation by potassium. Usually, large values of σ_D are found for the depolarization of resonance radiation by atoms of the same kind as the emitting atom.

The extremely large value of σ_D for the case of the depolarization of sodium resonance radiation by potassium was of unusual interest to the chemist at the time the work was published. The chemist had been looking for evidence of the transfer of energy by collision over large distances in order to account for the rates of first order monomolecular reactions. That the large value of σ_D found by Datta is not really due to a transfer of energy by a kinetic theory collision was shown by Foote[15]. He noticed that most atoms such as argon do not lead to large depolarizing diameters for sodium resonance radiation, and that these atoms have closed outer electronic shells and consequently no magnetic moment. Potassium, on the other hand, has only one valence electron in the outer shell and consequently shows a magnetic moment. From this remark of Foote's, it is quite easy to explain the large depolarizing effect of potassium. Thus, from the fact that it has a magnetic moment, potassium acts like a small magnet and the magnetic field due to the potassium atom is quite considerable at small distances (10^{-5} to 10^{-8} cm.). With a zero applied field the sodium atom might find itself in the field of a potassium atom, which could have a random direction with respect to the electric vector of the exciting light and thus cause depolarization of the resonance radiation. With large applied magnetic fields the depolarizing effect of potassium on sodium resonance radiation decreases. This effect is undoubtedly due to the setting in of the Paschen-Back effect of hyperfine structure in sodium.

In the case of the depolarization of resonance radiation by atoms of the same kind as those emitting the radiation, two factors come into play: (1) the effect of imprisoned resonance radiation; and (2) the effect of collisions. It is known that increasing the vapour pressure of gas has a large depolarizing effect on its own resonance radiation. In the past this has been ascribed to a large effective cross-section associated with depolarizing collisions. In view of the unknown effect of imprisoned resonance radiation it is unwise at this time to make any definite statements until further experimental work is done.

11. EFFECT OF ELECTRIC FIELDS ON RESONANCE RADIATION

11a. MEASUREMENTS ON FREQUENCY (STARK EFFECT). Two major accomplishments in the development of physics have been the discovery and explanation of the Zeeman effect and the Stark effect. The former has been extensively discussed in the preceding sections, and is concerned with the behaviour of emission and absorption lines in magnetic fields. Somewhat similar phenomena take place in strong electric fields, as was shown originally by Stark. The original experiment consisted in showing that an emission line of an element splits into several components if the emitting source is placed in a strong electric field, the separation and number of the various components depending on the line in question, and the strength and direction of the electric field. For light atoms, the separation of the Stark effect components is measurable with apparatus of ordinary resolving power. In the case of heavy atoms, such as sodium or mercury, the splitting is too small to be measured with the usual spectroscopic apparatus, even in the highest fields obtainable. For this reason indirect methods of measurement have been devised which depend on the properties of absorption lines discussed in Chap. III.

Ladenburg[27] investigated the effect of electric fields up to 160,000 volts/cm. on the absorption of the sodium D lines by sodium vapour. Light from a discharge tube containing sodium and an inert gas was sent through an absorption cell containing sodium vapour, which could be placed in a strong electric field, and was examined visually with a Lummer-Gehrcke plate and spectroscope. The light source was operated in such a way that the light intensity from it was continuous and uniform over each D line (breadth of line about 0·13 Å.), i.e. it exhibited neither self-reversal nor hyperfine structure. Observations on the D_2 line with the absorption cell in place, but in zero applied field, showed an absorption minimum at the centre of the line. In fields of from 95,000 to 160,000 volts/cm. the absorption minimum shifted toward the red by an amount between 0·009 Å. and 0·025 Å., increasing approximately as the square of the field strength.

An attempt at a quantitative measurement of the Stark effect broadening of the 2537 line of mercury was made by Brazdziunas[1], who used the method of comparing the form of an emission line from a resonance lamp with that of the absorption line in an absorption vessel (see Chap. III, §4h). A mercury arc excited radiation in a resonance lamp, and the light emitted therefrom was sent through an absorption cell, containing mercury vapour, and its intensity measured by means of a photoelectric cell. The resonance lamp was fitted with brass plates, to which could be applied a potential sufficient to give a field of 160,000 volts/cm. The experiment consisted in measuring the change of intensity of the resonance radiation transmitted through the absorption cell as a function of electric field applied on the resonance lamp, and comparing this with results obtained by applying a magnetic field of known strength to the resonance lamp. Taking the Zeeman splitting as known, and neglecting complications due to hyperfine structure, the Stark effect splitting was calculated. Brazdziunas found a shift of $5 \cdot 4 \times 10^{-4}$ Å. for the σ components (those polarized at right angles to the electric field) in a field of 100,000 volts/cm., and of $1 \cdot 9 \times 10^{-4}$ Å. for the π component in a field of 140,000 volts/cm. He further showed that the wavelength shift was proportional to the square of the applied electric field strength.

The results of the experiment are open to the criticism that the hyperfine structure of the 2537 line was not considered in making the calculation. Since the splitting due to Stark effect is less than the hyperfine structure splitting, and since the calculation was based on the assumption that the 2537 line splits into one π and two σ components in a magnetic field, the absolute values of the wave-length shifts obtained may be in error.

The Stark effect of some of the higher states of mercury was investigated by Terenin[46]. He investigated the intensity of the stepwise radiation produced in mercury vapour by strong illumination as a function of an electric field applied to the resonance vessel. He found that the intensity of certain lines, coming from higher states, decreased when the electric field was applied to the cell. The explanation is, of course, that the position of a given absorption line of the vapour is displaced by the field to such an extent that excitation by the corresponding line from the arc is impossible, the intensity of the fluorescent line being thereby decreased. The lines showing this effect were 5770, 5790, 3650, 3655, 3663, 3126, 3131 and 2967.

11*b*. MEASUREMENTS ON POLARIZATION. Hanle[21] made an investigation of the effect of electric fields on the polarization of mercury resonance radiation. The resonance vessel consisted of a glass bulb into which were fitted two brass plates, which, when charged by an electrostatic machine, were capable of producing an electric field, in the region between them, of 100,000 volts/cm. The vessel was equipped with suitable quartz windows so that mercury vapour contained therein could be excited and the resonance radiation observed.

With the incident light progressing in the Z direction, and the electric field parallel to X (see Fig. 70), the following experiments were made:

1. Observation parallel to Y.

(a) Electric vector parallel to X. The resonance radiation was linearly polarized parallel to X independent of the presence of the electric field.

(b) Electric vector parallel to Y. The radiation was weak and unpolarized in the absence of a field; strong and polarized parallel to Z in the presence of the field.

2. Observation direction at an angle of 25° to the incident beam (in YZ plane).

(a) Electric vector parallel to X. The resonance radiation polarized parallel to X independent of the presence of the field, as in 1 (a).

(b) Electric vector parallel to Y. Resonance radiation polarized parallel to Y in zero field, and slightly strengthened by the presence of the field.

(c) Electric vector at an angle of 45° to the X axis. In zero field the resonance radiation was polarized in a direction parallel to the incident electric vector. In the presence of a field of 100,000 volts/cm. the radiation was entirely unpolarized. In weaker fields the radiation was elliptically polarized.

The explanation of these experiments may be attempted on the basis of the Stark effect diagram for mercury (Fig. 84). Neglecting complications which may arise from hyperfine structure, the $6\,^3P_1$ state of mercury splits into three levels in an electric field: viz. $n_3 = 0$; $n_3 = \pm 1$; these last two coincide, however. From these states π and σ components may be produced by transitions to the $6\,^1S_0$ state.

If the electric vector is parallel to the field, a mercury atom will absorb only the π component reaching the upper state $n_3 = 0$, and will hence radiate a π component, so that the resonance radiation will be polarized as in experiment 1 (a).

Experiment 1 (*b*) is definitely not in agreement with the theory. In this case, the electric vector of the incident light is perpendicular to the electric field, so that only σ components are absorbed, and the states $n_3 = \pm 1$ will be reached. These states have the same energy, but the radiation given off when the atom drops back to the ground state will be circularly polarized about the field in either a right-handed or left-handed sense depending on the sign of n_3. In the language of quantum mechanics we may say that, on the average, the atom has as good a chance of being in one of the two states as the other. The radiation emitted should then consist of two circularly

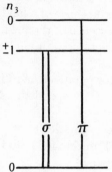

Fig. 84. Stark effect for 2537.

polarized components of the *same frequency* and *equal intensity*, and these two components should be capable of interfering with each other. If one observes the resonance radiation, as in experiment 1 (*b*), at right angles to the direction of the electric field, the theory predicts that the radiation should have exactly the same appearance as regards *intensity* and *polarization* as if no field were applied. This is in disagreement with experiment, which showed that both the intensity and the polarization increased on applying the field, the direction of polarization being parallel to Z.

A similar experiment was performed on sodium resonance radiation by Winkler[52], who found the same type of effect as that found by Hanle, which in this case is again in disagreement with theory.

The reason for the discrepancy between theory and experiment is possibly that the theory may have to be altered to take nuclear spin into consideration. It is to be hoped that this point will be investigated in the near future.

REFERENCES TO CHAPTER V

[1] Brazdziunas, P., *Ann. d. Phys.* **6**, 739 (1930).
[2] Breit, G., *Journ. Opt. Soc. Amer.* **10**, 439 (1925).
[3] —— *ibid.* **11**, 465 (1925).
[4] —— *Rev. Modern Phys.* **5**, 91 (1933).
[5] Breit, G. and Ellett, A., *Phys. Rev.* **25**, 888 (1925).
[6] Datta, G. L., *Z. f. Phys.* **37**, 625 (1926).
[7] Dirac, P. A. M., *Quantum Mechanics*, Oxford University Press.
[8] Ellett, A., *Nature*, **114**, 931 (1924).
[9] —— *Journ. Opt. Soc. Amer.* **10**, 427 (1925).
[10] —— *Phys. Rev.* **35**, 588 (1930).
[11] —— *ibid.* **37**, 216 (1931).
[12] Ellett, A. and Larrick, L., *ibid.* **39**, 294 (1932).
[13] Ellett, A. and MacNair, W. A., *ibid.* **31**, 180 (1928).
[14] Fermi, E. and Rasetti, F., *Z. f. Phys.* **33**, 246 (1925).
[15] Foote, P. D., *Phys. Rev.* **30**, 300 (1927).
[16] Gaviola, E. and Pringsheim, P., *Z. f. Phys.* **34**, 1 (1925).
[17] Granath, L. P. and Van Atta, C. M., *Phys. Rev.* **44**, 60 (1933).
[18] Gülke, R., *Z. f. Phys.* **56**, 524 (1929).
[19] Hanle, W., *Naturwiss.* **11**, 691 (1923).
[20] —— *Z. f. Phys.* **30**, 93 (1924); *Ergeb. der Exakten Naturwiss.* **4**, 214 (1925).
[21] —— *Z. f. Phys.* **35**, 346 (1926).
[22] —— *ibid.* **41**, 164 (1927).
[23] Hanle, W. and Richter, E. F., *ibid.* **54**, 811 (1929).
[24] Heisenberg, W., *ibid.* **31**, 617 (1926).
[25] Heydenburg, N. P., *Phys. Rev.* **43**, 640 (1933).
[26] Heydenburg, N. P., Larrick, L. and Ellett, A., *ibid.* **40**, 1041 (1932).
[27] Ladenburg, R., *Z. f. Phys.* **28**, 31 (1924).
[28] Larrick, L. (Thesis).
[29] Larrick, L. and Heydenburg, N. P., *Phys. Rev.* **39**, 289 (1932).
[30] MacNair, W. A., *ibid.* **29**, 766 (1927).
[31] —— *Proc. Nat. Acad. Sci.* **13**, 430 (1927).
[32] Mitchell, A. C. G., *Phys. Rev.* **36**, 1589 (1930).
[33] —— *ibid.* **38**, 473 (1931).
[34] —— *ibid.* **40**, 964 (1932).
[35] —— *ibid.* **43**, 887 (1933).
[36] Mrozowski, S., *Bull. Acad. Pol. Sci.* (1930 and 1931), No. 6 A, p. 489.
[37] Olson, H. F., *Phys. Rev.* **32**, 443 (1928).
[38] Pringsheim, P. and Gaviola, E., *Z. f. Phys.* **25**, 690 (1924).
[39] Rabi, I. I. and Cohen, V., *Phys. Rev.* **43**, 582 (1933).

[39 a] Randall, R. H., *Phys. Rev.* **35**, 1161 (1930).
[40] Rayleigh, *Proc. Roy. Soc.* **102**, 190 (1922).
[41] Richter, E. F., *Ann. d. Phys.* **7**, 293 (1930).
[42] Schüler, H. and Brück, H., *Z. f. Phys.* **55**, 575 (1929).
[43] Schüler, H. and Keyston, J., *ibid.* **67**, 433 (1931).
[44] Soleillet, P., *Compt. Rend.* **185**, 198 (1927); **187**, 212 (1928).
[45] —— *ibid.* **187**, 723 (1928).
[46] Terenin, A., *Z. f. Phys.* **37**, 676 (1926).
[47] Urey, H. C. and Joffe, J., *Phys. Rev.* **43**, 761 (1933).
[48] Van Vleck, J. H., *Proc. Nat. Acad. Sci.* **11**, 612 (1925).
[49] Von Keussler, V., *Ann. d. Phys.* **87**, 793 (1927).
[50] —— *Z. f. Phys.* **73**, 565 (1932).
[51] Weisskopf, V., *Ann. d. Phys.* **9**, 23 (1931).
[52] Winkler, E., *Z. f. Phys.* **64**, 799 (1930).
[53] Wood, R. W., *Phil. Mag.* **44**, 1109 (1922).
[54] Wood, R. W. and Ellett, A., *Proc. Roy. Soc.* **103**, 396 (1923); *Phys. Rev.* **24**, 243 (1924).

APPENDIX

I. ABSORPTION COEFFICIENT OF A GAS. On the basis of the electron theory of dispersion, Voigt [Ref. 75, Chap. III] showed that the absorption coefficient of a gas, when Doppler effect and natural damping are present, is given by

$$n\kappa = \frac{\sqrt{\pi}\,e^2 Nf}{m\omega_0 bn} \int_0^\infty 2y\, e^{-y^2} \arctan \frac{b\nu' y}{\left(\dfrac{\nu'}{2}\right)^2 + \mu^2 - b^2 y^2}\, dy \quad \ldots(209),$$

where $n =$ index of refraction,

         $n\kappa =$ electron theory absorption coefficient,

         $\omega_0 =$ frequency at the centre of the line,

         $b = \dfrac{\omega_0}{c} \sqrt{\dfrac{2RT}{M}}$,

         $\nu' =$ atomic damping constant,

         $\mu =$ frequency distance from centre of line.

In the notation of this book, the above quantities are as follows:

$$n = 1,$$

$$n\kappa = \frac{\lambda_0}{4\pi} k_\nu,$$

$$\omega_0 = 2\pi\nu_0,$$

$$b = \frac{\pi}{\sqrt{\ln 2}} \Delta\nu_D,$$

$$\nu' = 2\pi\Delta\nu_N \left(= \frac{1}{\tau}\right),$$

$$\mu = 2\pi(\nu - \nu_0),$$

from which it is evident that

$$\frac{\nu'}{2b} = \frac{\Delta\nu_N}{\Delta\nu_D} \sqrt{\ln 2} = a \quad \text{[see Eq. (39)]},$$

and $$\frac{\mu}{b} = \frac{2(\nu - \nu_0)}{\Delta\nu_D} \sqrt{\ln 2} = \omega \quad \text{[see Eq. (38)]}.$$

In the notation of this book, Voigt's formula then becomes

$$k_\nu = \frac{1}{\pi} \frac{2}{\Delta\nu_D} \sqrt{\frac{\ln 2}{\pi}} \cdot \frac{\pi e^2 Nf}{mc} \int_0^\infty 2ye^{-y^2} \arctan \frac{2ay}{a^2 + \omega^2 - y^2} \, dy,$$

and in virtue of Eq. (35), namely

$$k_0 = \frac{2}{\Delta\nu_D} \sqrt{\frac{\ln 2}{\pi}} \cdot \frac{\pi e^2 Nf}{mc},$$

Voigt's formula becomes finally

$$k_\nu = \frac{k_0}{\pi} \int_0^\infty 2ye^{-y^2} \arctan \frac{2ay}{a^2 + \omega^2 - y^2} \, dy.$$

Now it can easily be shown that

$$\arctan \frac{2ay}{a^2 + \omega^2 - y^2} = \arctan \frac{\omega + y}{a} - \arctan \frac{\omega - y}{a},$$

whence, Voigt's formula becomes

$$k_\nu = \frac{k_0}{\pi} \left[\int_0^\infty 2ye^{-y^2} \arctan \frac{\omega + y}{a} \, dy - \int_0^\infty 2ye^{-y^2} \arctan \frac{\omega - y}{a} \, dy \right]$$
$$\cdots\cdots(210),$$

and upon integrating by parts

$$k_\nu = \frac{k_0}{\pi} \left[\int_0^\infty e^{-y^2} \frac{\frac{1}{a} dy}{1 + \left(\frac{\omega + y}{a}\right)^2} + \int_0^\infty e^{-y^2} \frac{\frac{1}{a} dy}{1 + \left(\frac{\omega - y}{a}\right)^2} \right]$$

$$= k_0 \frac{a}{\pi} \int_{-\infty}^\infty \frac{e^{-y^2} dy}{a^2 + (\omega - y)^2} \qquad \cdots\cdots(211),$$

which is identical with Eq. (40).

Now $$\frac{a}{a^2 + (\omega - y)^2} = \int_0^\infty e^{-ax} \cos\left[(\omega - y)x\right] dx.$$

Therefore

$$k_\nu = \frac{k_0}{\pi} \int_0^\infty e^{-ax} dx \int_{-\infty}^\infty e^{-y^2} \cos\left[(\omega - y)x\right] dy$$

$$= \frac{k_0}{\pi} \int_0^\infty e^{-ax} \cos \omega x \, dx \int_{-\infty}^\infty e^{-y^2} \cos xy \, dy,$$

and since $$\int_{-\infty}^\infty e^{-y^2} \cos xy \, dy = \sqrt{\pi} e^{-\frac{x^2}{4}},$$

$$k_\nu = \frac{k_0}{\sqrt{\pi}} \int_0^\infty e^{-ax - \frac{x^2}{4}} \cos \omega x \, dx \qquad \cdots\cdots(212),$$

which is the form given by Reiche [Ref. 60, Chap. III].

II. Value of $\dfrac{a}{\pi}\displaystyle\int_{-\infty}^{\infty}\dfrac{e^{-y^2}dy}{a^2+(\omega-y)^2}$ for small values of a.

Using Reiche's form of this integral and assuming that $a \leqslant 0\cdot 01$,

$$\frac{k_\nu}{k_0} = \frac{1}{\sqrt{\pi}}\int_0^\infty e^{-\frac{x^2}{4}}\cos\omega x\,dx - \frac{a}{\sqrt{\pi}}\int_0^\infty xe^{-\frac{x^2}{4}}\cos\omega x\,dx$$

$$= \frac{2}{\sqrt{\pi}}\int_0^\infty e^{-t^2}\cos 2\omega t\,dt - \frac{2a}{\sqrt{\pi}}\int_0^\infty 2te^{-t^2}\cos 2\omega t\,dt$$

$$= \frac{2}{\sqrt{\pi}}\int_0^\infty e^{-t^2}\cos 2\omega t\,dt - \frac{2a}{\sqrt{\pi}}\left[1 - 2\omega\int_0^\infty e^{-t^2}\sin 2\omega t\,dt\right].$$

Using the formulas

$$\int_0^\infty e^{-t^2}\cos 2\omega t\,dt = \frac{\sqrt{\pi}}{2}e^{-\omega^2}$$

and $$\int_0^\infty e^{-t^2}\sin 2\omega t\,dt = e^{-\omega^2}\int_0^\omega e^{x^2}dx = F(\omega),$$

the absorption coefficient assumes the form given by Eq. (41), namely

$$\frac{k_\nu}{k_0} = e^{-\omega^2} - \frac{2a}{\sqrt{\pi}}[1 - 2\omega F(\omega)] \qquad \ldots\ldots(213).$$

The following table of values of $F(\omega)$ was obtained from the very complete table of W. Lash Miller and A. R. Gordon in the *Journal of Physical Chemistry*, **35**, 2878 (1931).

It is interesting to note that

$$1 - 2\omega F(\omega) = \frac{d}{d\omega}F(\omega).$$

For small values of ω

$$-2\omega F(\omega) = 1 - \frac{2\omega^2}{1} + \frac{(2\omega^2)^2}{1.3} - \frac{(2\omega^2)^3}{1.3.5} + \frac{(2\omega^2)^4}{1.3.5.7} - \ldots \quad \ldots(214),$$

whereas for large values of ω

$$1 - 2\omega F(\omega) = -\left[\frac{1}{2\omega^2} + \frac{1.3}{(2\omega^2)^2} + \frac{1.3.5}{(2\omega^2)^3} + \frac{1.3.5.7}{(2\omega^2)^4} + \ldots\right]\ldots(215).$$

ω	$F(\omega)$	$1-2\omega F(\omega)$	ω	$F(\omega)$	$1-2\omega F(\omega)$
·0	0·0000	1·0000	6·0	·08454	− ·01451
·2	·1948	·9221	6·2	·08174	− ·01355
·4	·3599	·7121	6·4	·07912	− ·01268
·6	·4748	·4303	6·6	·07666	− ·01190
·8	·5321	·1487	6·8	·07435	− ·01118
1·0	·5381	− ·07616	7·0	·07218	− ·01053
1·2	·5073	− ·2175	7·2	·07013	− ·009938
1·4	·4565	− ·2782	7·4	·06820	− ·009393
1·6	·3999	− ·2797	7·6	·06637	− ·008892
1·8	·3468	− ·2485	7·8	·06464	− ·008429
2·0	·3013	− ·2052	8·0	·06300	− ·008000
2·2	·2645	− ·1638	8·2	·06144	− ·007608
2·4	·2353	− ·1295	8·4	·05995	− ·007242
2·6	·2122	− ·1033	8·6	·05854	− ·006902
2·8	·1936	− ·08389	8·8	·05719	− ·006586
3·0	·1783	− ·06962	9·0	·05591	− ·006290
3·2	·1655	− ·05896	9·2	·05467	− ·006014
3·4	·1545	− ·05076	9·4	·05350	− ·005757
3·6	·1450	− ·04430	9·6	·05237	− ·005516
3·8	·1367	− ·03908	9·8	·05129	− ·005290
4·0	·1293	− ·03480	10·0	·05025	− ·005076
4·2	·1228	− ·03119	10·2	·04926	− ·004877
4·4	·1168	− ·02815	10·4	·04830	− ·004688
4·6	·1115	− ·02554	10·6	·04738	− ·004511
4·8	·1066	− ·02336	10·8	·04650	− ·004344
5·0	·1021	− ·02134	11·0	·04564	− ·004183
5·2	·09804	− ·01963	11·2	·04482	− ·004035
5·4	·09427	− ·01812	11·4	·04403	− ·003893
5·6	·09078	− ·01678	11·6	·04327	− ·003757
5·8	·08755	− ·01558	11·8	·04253	− ·003630
			12·0	·04181	− ·003510

III. LINE ABSORPTION A_L. The line absorption A_L is defined as [see Eq. (58)]

$$A_L = \frac{\int_{-\infty}^{\infty} (1 - e^{-k_0 l e^{-\omega^2}})^2 \, d\omega}{\int_{-\infty}^{\infty} (1 - e^{-k_0 l e^{-\omega^2}}) \, d\omega},$$

and is evaluated by means of the series

$$A_L = \frac{a_1 k_0 l - a_2 (k_0 l)^2 + \ldots + a_n (k_0 l)^n - \ldots}{1 - b_1 k_0 l + b_2 (k_0 l)^2 - \ldots - b_n (k_0 l)^n + \ldots},$$

where $\quad a_n = \dfrac{2^{n+1} - 2}{(n+1)! \sqrt{n+1}}, \qquad b_n = \dfrac{1}{(n+1)! \sqrt{n+1}}.$

The following table was taken from the papers of H. Kopfermann and W. Tietze, *Z. f. Phys.* **56**, 604 (1929), and R. Ladenburg and S. Levy, *ibid.* **65**, 189 (1930).

$k_0 l$	A_L	$k_0 l$	A_L	$k_0 l$	A_L
0·1	·070	1·1	·491	2·1	·675
0·2	·129	1·2	·516	2·2	·685
0·3	·181	1·3	·538	2·3	·695
0·4	·232	1·4	·562	2·4	·706
0·5	·284	1·5	·583	2·5	·715
0·6	·327	1·6	·602	2·6	·724
0·7	·366	1·7	·619	2·7	·732
0·8	·401	1·8	·634	2·8	·738
0·9	·433	1·9	·649	3·0	·750
1·0	·465	2·0	·662	4·0	·800
				5·0	·835

IV. THE ABSORPTION A_α. This is defined as [see Eq. (61)]

$$A_\alpha = \frac{\displaystyle\int_{-\infty}^{\infty} e^{-\left(\frac{\omega}{\alpha}\right)^2}(1 - e^{-k_0 l e^{-\omega^2}})\, d\omega}{\displaystyle\int_{-\infty}^{\infty} e^{-\left(\frac{\omega}{\alpha}\right)^2}\, d\omega}$$

$$= \frac{k_0 l}{\sqrt{1+\alpha^2}} - \frac{(k_0 l)^2}{2!\sqrt{1+2\alpha^2}} + \ldots + (-1)^n \frac{(k_0 l)^n}{n!\sqrt{1+n\alpha^2}} + \ldots.$$

The following table is the result partly of the use of the above series, partly of graphical integration and partly of graphical interpolation.

$\dfrac{\Delta \nu_E}{\Delta \nu_D} = a$	0	0·5	1·0	1·5	2·0	2·5	3·0
$k_0 l$							
0	0	0	0	0	0	0	0
·25	·221	·200	·160	·125	·102	·086	·0723
·50	·393	·360	·291	·229	·188	·159	·133
1·0	·632	·588	·486	·385	·316	·265	·226
1·5	·777	·736	·619	·494	·400	·336	·287
2·0	·865	·832	·711	·575	·472	·400	·348
3·0	·950	·925	·820	·674	·564	·476	·414
4·0	·982	·967	·878	·740	·622	·532	·461
4·5	·989	·977	·897	·762	·640	·549	·480

V. The Function S. This is defined as [see § 4f, Chap. III]

$$S = \frac{1}{\sqrt{\pi} k_0 l} \int_{-\infty}^{\infty} (1 - e^{-k_0 l e^{-\omega^2}}) \, d\omega$$

$$= 1 - \frac{k_0 l}{2! \sqrt{2}} + \frac{(k_0 l)^2}{3! \sqrt{3}} - \frac{(k_0 l)^3}{4! \sqrt{4}} + \dots$$

The following table of values of S was obtained from the paper of R. Ladenburg and S. Levy, *Z. f. Phys.* **65**, 189 (1930). A convenient relation exists between S and A_L, namely

$$A_L = 2 - \frac{2S(2k_0 l)}{S(k_0 l)}.$$

$k_0 l$	S	$k_0 l S$	$k_0 l$	S	$k_0 l S$
0	1·000	0	1·2	·683	·820
·10	·964	·0964	1·4	·646	·905
·15	·948	·142	1·6	·616	·985
·20	·933	·187	1·8	·584	1·050
·25	·917	·229	2·0	·556	1·112
·30	·902	·270	2·2	·532	1·170
·35	·887	·311	2·4	·507	1·218
·40	·872	·348	2·6	·487	1·267
·45	·859	·387	2·8	·468	1·311
·50	·844	·421	3·0	·450	1·350
·55	·831	·457	3·2	·432	1·385
·60	·818	·491	3·4	·417	1·415
·65	·806	·524	4·0	·372	1·488
·70	·793	·555	4·4	·347	1·530
·75	·780	·585	5	·316	1·580
·80	·768	·620	6	·276	1·656
·85	·757	·640	7	·246	1·720
·90	·745	·675	8	·222	1·778
·95	·734	·700	9	·202	1·820
1·00	·725	·725	10	·186	1·860

VI. The Absorption $A'_{k_0' l'}$. This is defined as [see Eq. (59)]

$$A'_{k_0' l'} = \frac{\int_{-\infty}^{\infty} (1 - e^{-k_0' l' e^{-\omega^2}})(1 - e^{-k_0 l e^{-\omega^2}}) \, d\omega}{\int_{-\infty}^{\infty} (1 - e^{-k_0' l' e^{-\omega^2}}) \, d\omega}.$$

From the definition of the function S, it is apparent that

$$A'_{k_0' l'} = \frac{1}{\sqrt{\pi} k_0' l' \, S(k_0' l')} \int_{-\infty}^{\infty} (1 - e^{-k_0' l' e^{-\omega^2}})(1 - e^{-k_0 l e^{-\omega^2}}) \, d\omega,$$

whence expanding the expression $(1 - e^{-k_0'l'e^{-\omega^2}})$ in a series and making use of the function A_α, we get

$$A'_{k_0'l'} = \frac{1}{S(k_0'l')}\left[A_1(k_0l) - \frac{k_0'l'}{2!\sqrt{2}}A_{1/\sqrt{2}}(k_0l) \right.$$
$$\left. + \frac{(k_0'l')^2}{3!\sqrt{3}}A_{1/\sqrt{3}}(k_0l) - \frac{(k_0'l')^3}{4!\sqrt{4}}A_{1/\sqrt{4}}(k_0l) + \dots \right].$$

The following table was obtained from the above series and with the aid of the reciprocal relation

$$x\,S(x)\,A_x'(y) = y\,S(y)\,A_y'(x).$$

VALUES OF $A'_{k_0'l'}$.

k_0l \ $k_0'l'$	0	·5	1·0	1·5	2·0	2·5	3·0	4·0
0	0	0	0	0	0	0	0	0
·25	·160	·156	·153	·149	·147	·145	·144	·135
·50	·291	·285	·279	·273	·267	·262	·258	·246
1·0	·486	·475	·465	·454	·445	·437	·429	·419
1·5	·619	·608	·597	·583	·573	·562	·551	·540
2·0	·711	·700	·687	·674	·662	·651	·642	·626
3·0	·820	·811	·800	·789	·776	·764	·750	·740
4·0	·878	·870	·861	·850	·839	·828	·817	·800
4·5	·897	·889	·880	·869	·858	·848	·839	—

VII. KUHN'S THEORY OF MAGNETO-ROTATION. [Ref. 32, Chap. III.] According to the classical dispersion theory, and in agreement with the quantum theory, the index of refraction n of a gas at the frequency ν in the neighbourhood of an absorption line at the frequency ν_0 is given by [see Eq. (87)]

$$n - 1 = \frac{e^2 Nf}{4\pi m\nu_0(\nu - \nu_0)}.$$

If the gas be placed in a magnetic field and be traversed by a beam of plane polarized light travelling in the direction of the lines of force, there will result a rotation of the plane of polarization, because of the difference in magnitude between the index of refraction for right-handed circularly polarized light and that for left-handed.

In order to calculate the index of refraction for right-handed circularly polarized light travelling parallel to a magnetic field of strength H, it is necessary to take into account that, in place of the undisturbed absorption line at ν_0, there are various right-handed circularly polarized Zeeman components with intensities β_s and at frequencies $\nu_0 + \alpha\alpha_s$, where the subscript s refers to the particular Zeeman component, α represents the normal Zeeman separation

$$\alpha = \frac{He}{4\pi mc},$$

and α_s the splitting factor of the sth component. The index of refraction for right-handed circularly polarized light at the frequency ν is then given by

$$n_- - 1 = \frac{Nfe^2}{4\pi m\nu_0} \sum_s \frac{\beta_s}{\nu - \nu_0 + \alpha\alpha_s}.$$

From the principle of spectroscopic stability, the β_s or relative intensities of the various Zeeman components satisfy the condition that

$$\Sigma\beta_s = 1.$$

For left-handed circularly polarized light at the frequency ν, the index of refraction is given by

$$n_+ - 1 = \frac{Nfe^2}{4\pi m\nu_0} \sum_s \frac{\beta_s}{\nu - \nu_0 - \alpha\alpha_s}.$$

If χ_ν denote the rotation of the plane of polarized light in traversing a layer of gas of thickness l, then

$$\frac{\chi_\nu}{l} = \frac{\pi\nu_0}{c}(n_- - n_+) = \frac{Nfe^2}{4mc} \sum_s \left(\frac{\beta_s}{\nu - \nu_0 + \alpha\alpha_s} - \frac{\beta_s}{\nu - \nu_0 - \alpha\alpha_s} \right)$$

$$= -\frac{Nfe^2\alpha}{2mc} \sum_s \frac{\alpha_s\beta_s}{(\nu - \nu_0)^2 - \alpha^2\alpha_s^2}.$$

If we limit ourselves to the edges of the absorption line, so that

$$(\nu - \nu_0)^2 \gg \alpha^2\alpha_s^2,$$

then we can write

$$\frac{\chi_\nu}{l} = -\frac{Nfe^2\alpha}{2mc(\nu - \nu_0)^2} \Sigma_s \alpha_s\beta_s,$$

and introducing

$$\alpha = \frac{He}{4\pi mc},$$

and

$$\mu = 2\pi(\nu - \nu_0),$$

there results
$$\chi_\nu = \frac{\pi e^3 Hl}{2m^2 c^2} \cdot \frac{Nf}{\mu^2} \sum_s \alpha_s \beta_s,$$

which in comparison with Eq. (72) shows that

$$z = \sum_s \alpha_s \beta_s \qquad \ldots\ldots(216).$$

The quantity z indicates how much bigger the observed magneto-rotation will be in the neighbourhood of an absorption line with anomalous Zeeman effect (characterized by the splitting factors α_s and the intensities β_s of the circularly polarized components) than the magneto-rotation in the neighbourhood of a line of the same intensity but with normal Zeeman effect. Since the splitting factors and relative intensities of the Zeeman components of all normal multiplets are known, z can be easily computed. They are given in the following table:

Line	α_1	β_1	α_2	β_2	$z = \sum_s \alpha_s \beta_s$
$1\,^1S_0 - 2\,^1P_1$	1	1	—	—	1
$1\,^1S_0 - 2\,^3P_1$	$\frac{3}{2}$	1	—	—	$\frac{3}{2}$
$1\,^2S_{1/2} - 2\,^2P_{1/2}$	$\frac{4}{3}$	1	—	—	$\frac{4}{3}$
$1\,^2S_{1/2} - 2\,^2P_{3/2}$	1	$\frac{3}{4}$	$\frac{5}{3}$	$\frac{1}{4}$	$\frac{7}{6}$
$2\,^2P_{1/2} - 2\,^2S_{1/2}$	$\frac{4}{3}$	1	—	—	$\frac{4}{3}$
$2\,^2P_{1/2} - 3\,^2D_{3/2}$	$\frac{13}{15}$	$\frac{3}{4}$	$\frac{11}{15}$	$\frac{1}{4}$	$\frac{5}{6}$

The values of α_s in the above table are strictly accurate only for values of the field strength H in the neighbourhood of 1000 gauss or more. The experiments of Minkowski show, however, in the case of the sodium D lines that no error is introduced at 300 gauss, and even as low as 30 gauss a difference of only 5 per cent. was noted by Weingeroff. This is due to the fact that, in the neighbourhood of 300 gauss, the Paschen-Back effect of the hyperfine-structure components sets in [see Chap. v].

VIII. EFFECT OF HYPERFINE STRUCTURE ON THE VALUE OF χ_ν. It was shown by Weingeroff [Ref. 81, Chap. III] that when a spectral line consists of hyperfine-structure components of which the ith component is at the frequency ν_i, the magneto-rotation may be calculated by the method of Kuhn, provided one sums over all Zeeman components of all hyperfine-structure components. Therefore

$$n_\pm - 1 = \sum_i \frac{Nf_i e^2}{4\pi m\nu_0} \sum_{s_i} \frac{\beta_{s_i}}{\nu - \nu_i \mp \alpha\alpha_{s_i}},$$

and the magneto-rotation is

$$\chi_\nu = -\sum_i \frac{Nf_i e^2 \alpha}{2mc} \sum_{s_i} \frac{\alpha_{s_i}\beta_{s_i}}{(\nu - \nu_i)^2 - \alpha^2\alpha_{s_i}^2}$$

At the edges of any component where

$$(\nu - \nu_i)^2 \gg \alpha^2\alpha_{s_i}^2,$$

we have

$$\chi_\nu = -\sum_i \frac{Nf_i e^2 \alpha}{2mc\,(\nu - \nu_i)^2} \sum_{s_i} \alpha_{s_i}\beta_{s_i},$$

and finally, if we denote by ν_0 the frequency of the centre of gravity of the hyperfine-structure components, we can write approximately

$$\chi_\nu = -\frac{\pi e^3 Hl}{2m^2 c^2} \cdot \frac{Nf}{\mu^2} \sum_i K_i \sum_{s_i} \alpha_{s_i}\beta_{s_i},$$

where K_i is defined by the formula

$$f_i = K_i f$$

and must satisfy the condition that

$$\Sigma K_i = 1.$$

In comparison with Eq. (72), it is apparent that, in this case,

$$z = \sum_i K_i \sum_{s_i} \alpha_{s_i}\beta_{s_i} \qquad \ldots\ldots(217),$$

which shows that z depends not only on the relative intensities β_{s_i} and splitting factors α_{s_i} of the Zeeman components, but also on the relative f-values of the hyperfine-structure components.

IX. VALUE OF $\dfrac{a'}{\pi} \displaystyle\int_{-\infty}^{\infty} \dfrac{e^{-v^2}dy}{a'^2 + (\omega - y)^2}$ FOR LARGE VALUES OF

a'. A series expansion was used to evaluate this integral for $a' = 0.5$, 1.0 and 1.5 according to a method due to T. H. Gronwall and given in a paper by M. W. Zemansky, *Phys. Rev.* **36**, 919 (1930). The values for $a' = 2$ and 10 were obtained from a table in Born's *Optik*, p. 486.

$a'=0$

ω	$\dfrac{k_\nu}{k_0}$
·0	1·0000
·2	·9608
·4	·8521
·6	·6977
·8	·5273
1·0	·3679
1·2	·2369
1·4	·1409
1·6	·0773
1·8	·0392
2·0	·0183
2·2	·0079
2·4	·0032
2·6	·0012
2·8	·0004
3·0	·0001

$a'=0.5$

ω	$\dfrac{k_\nu}{k_0}$
·0	·6157
·2	·6015
·4	·5613
·6	·5011
·8	·4294
1·0	·3549
1·2	·2846
1·4	·2233
1·6	·1728
1·8	·1333
2·0	·1034
4·0	·0183
6·0	·0081
8·0	·004
10·0	·003

$a'=1$

ω	$\dfrac{k_\nu}{k_0}$
·0	·4276
·2	·4215
·4	·4038
·6	·3766
·8	·3425
1·0	·3047
1·2	·2662
1·4	·2292
1·6	·1954
1·8	·1657
2·0	·1402
2·2	·120
2·4	·102
2·6	·088
2·8	·078
3·0	·066
3·2	·057
3·4	·051
3·6	·045
3·8	·041
4·0	·037
6·0	·016
8·0	·009
10·0	·005

$a'=1.5$

ω	$\dfrac{k_\nu}{k_0}$
·0	·3216
·2	·3186
·4	·3097
·6	·2958
·8	·2779
1·0	·2571
1·2	·2349
1·4	·2123
1·6	·1902
1·8	·1695
2·0	·1504
4·0	·0487
6·0	·0228
8·0	·0131
10·0	·0083

$a'=2$

ω	$\dfrac{k_\nu}{k_0}$
·0	·257
·4	·252
·8	·236
1·2	·212
1·6	·178
2·0	·148
2·4	·123
2·8	·101
3·2	·0850
3·6	·0708
4·0	·0598
4·4	·0505
4·8	·0440
5·2	·0378
5·6	·0330
6·0	·0291
6·4	·0259
6·8	·0231
7·2	·0208
7·6	·0186
8·0	·0169

$a'=10$

ω	$\dfrac{k_\nu}{k_0}$
0	·0561
2	·0541
4	·0486
6	·0414
8	·0344
10	·0283
12	·0232
14	·0191
16	·0159
18	·0134
20	·0114
22	·00965
24	·00835
26	·00728
28	·00637
30	·00564
32	·00502
34	·00451
36	·00406
38	·00366
40	·00333

X. DIFFUSED TRANSMITTED RESONANCE RADIATION.

When a collimated beam of resonance radiation of frequency between ν and $\nu + d\nu$ and intensity K is incident in the positive x direction upon the face $x = 0$ of a slab of gas whose absorption coefficient for this radiation is k, the intensity of the radiation emitted in the positive x direction from the face $x = l$ is given by

$$\pi I_+\,(x = l) = KG\,(kl, \tau Z_Q) \quad \text{[see Eq. (141)]},$$

where $G\,(kl, \tau Z_Q)$

$$= \frac{1}{2\,(1 - 3\tau Z_Q)} \left[\frac{6\sqrt{\tau Z_Q(1 + \tau Z_Q)} + e^{-kl}\sinh 2kl\sqrt{\dfrac{\tau Z_Q}{1 + \tau Z_Q}}}{\sinh\left(2kl\sqrt{\dfrac{\tau Z_Q}{1 + \tau Z_Q}} + 2\sinh^{-1}\sqrt{\tau Z_Q}\right)} - 3e^{-kl} \right],$$

where $\tau =$ lifetime of the excited state,

$Z_Q =$ number of quenching collisions per second per excited atom.

The following table of values of G was obtained from a paper by M. W. Zemansky, *Phys. Rev.* **36**, 919 (1930).

VALUES OF $G\,(kl, \tau Z_Q)$.

kl \ τZ_Q	·0	·05	·10	·20	·333	·50
0·5	·194	·175	·164	·143	·125	·107
1·0	·290	·260	·236	·198	·160	·128
1·5	·332	·282	·244	·195	·150	·118
2·0	·344	·273	·227	·168	·124	·092
2·5	·334	—	—	—	·0968	—
3·0	·320	·219	·163	·106	·0704	·0488
3·5	·300	—	—	—	·0504	—
4·0	·280	·158	·104	·0590	·0351	·0224
4·5	·260	—	—	—	·0241	—
5·0	·243	·108	·0628	·0308	·0163	·00968

XI. SAMSON'S EQUIVALENT OPACITY.

The equivalent opacity, $\bar{k}l$, is defined by Samson as follows [see Eq. (142)]:

$$e^{-\bar{k}l} = \frac{\displaystyle\int_{-\infty}^{\infty} e^{-\omega^2} \cdot e^{-k_0 l e^{-\omega^2}}\,d\omega}{\displaystyle\int_{-\infty}^{\infty} e^{-\omega^2}\,d\omega},$$

where l is the thickness of the absorbing layer. The following values were obtained by graphical integration.

k_0l	$\bar{k}l$
0	0
1	0·665
2	1·241
3	1·715
4	2·104
9·7	3·29
14·4	3·76
19·9	4·15

XII. KENTY'S EQUIVALENT OPACITY. The equivalent opacity $\bar{k}l$ is defined by Kenty as follows [see Eq. (158)]:

$$\bar{k}l = \sqrt{\frac{3}{4}} \sqrt{\frac{\pi}{2}} \cdot \sqrt{\frac{k_0 l}{\sqrt{2} F \sqrt{\ln k_0 l} - F \sqrt{2 \ln k_0 l}}},$$

where

$$F(\omega) = e^{-\omega^2} \int_0^\omega e^{x^2} dx.$$

The following table was obtained with the aid of the table of values of $F(\omega)$ given in Appendix II.

k_0l	$\bar{k}l$	k_0l	$\bar{k}l$
1·5	3·05	100	21·2
2	2·76	200	31·4
3	2·97	500	54·2
4	3·31	1000	77·8
5	3·70	2000	114
10	5·39	3000	142
15	6·85	4000	166
20	8·10	5000	186
30	10·4	6000	205
40	12·3	7000	223
50	14·1	8000	240

XIII. POLARIZATION OF RESONANCE RADIATION EXCITED BY UNPOLARIZED LIGHT. In making experiments on the polarization of resonance radiation, it is sometimes convenient to use an unpolarized exciting source, on account of the gain in light intensity thereby afforded, and to be able to convert this data into a form which will be comparable with the expressions developed in Chap. v. Suppose that a parallel beam of exciting light is progressing in the Z direction (Fig. 70) and that observations of the polarization of the resonance radiation are to be made along Y. In this case the electric vectors of the exciting light lie in the XY plane. Three interesting cases arise for computation.

(1) *Strong Magnetic Field in Direction of Incident Beam.* In this case the electric vectors all lie perpendicular to the field so that $\theta = \pi/2$, and the polarization is $P_\perp$.

(2) *Strong Magnetic Field at Right Angles to Incident Beam and Observation Direction.* With the field parallel to X we may resolve the incident light into two polarized components. One polarized parallel to the field $(I_\|)$ and one perpendicular to this $(I_\perp)$. To calculate the polarization, we write down the following equations, which follow from Eq. (182):

$$\xi(I_\|) = I_\| \, C \sum_\mu \frac{A_\pi{}^\mu}{A_\pi{}^\mu + A_\sigma{}^\mu} A_\pi{}^\mu,$$

$$\xi(I_\perp) = I_\perp \frac{C}{2} \sum_\mu \frac{A_\pi{}^\mu}{A_\pi{}^\mu + A_\sigma{}^\mu} A_\sigma{}^\mu,$$

$$\eta(I_\|) = I_\| \frac{C}{2} \sum_\mu \frac{A_\sigma{}^\mu}{A_\pi{}^\mu + A_\sigma{}^\mu} A_\pi{}^\mu,$$

$$\eta(I_\perp) = I_\perp \frac{C}{4} \sum_\mu \frac{A_\sigma{}^\mu}{A_\pi{}^\mu + A_\sigma{}^\mu} A_\sigma{}^\mu.$$

The polarization is now $\overline{P}$, defined by

$$\overline{P} = \frac{\xi(I_\|) + \xi(I_\perp) - \eta(I_\|) - \eta(I_\perp)}{\xi(I_\|) + \xi(I_\perp) + \eta(I_\|) + \eta(I_\perp)}.$$

Remembering that $I_\| = I_\perp$, and using Eq. (187), a short calculation shows that

$$\overline{P} = \frac{P_\|}{4 - P_\|} \qquad \dots\dots(218),$$

where $P_\|$ is the polarization observed with the incident light polarized parallel to the field H which is also parallel to X.

(3) *Zero Magnetic Field.* If the resonance tube is situated in a zero magnetic field, we must make an application of the Principle of Spectroscopic Stability. Let the incident beam be

resolved into two components polarized at right angles to each other, the one parallel to X, the other parallel to Y. Let their intensities be I_X and I_Y respectively. Consider first the resonance radiation due to excitation by the component polarized parallel to X. On account of spectroscopic stability we must assume, in this case, that the resonance tube is situated in a small magnetic field parallel to X. This will give rise to resonance radiation one component of which will be polarized parallel to X, and which we shall call $\xi_X(I_X)$. The other component will be circularly polarized about the field, but we shall see only the projection of this, since we observe in a direction perpendicular to the field. This radiation will be polarized along Z and will be called $\eta_Z(I_X)$. For excitation by I_Y, however, we must assume a small field parallel to the observation direction. The emitted resonance radiation will then consist of a part linearly polarized along Y, which will not be seen by the observer, and a part which will be circularly polarized about Y. This latter may now be resolved into two components of equal intensity polarized parallel to X and Z, respectively, which we may call $\eta_X(I_Y)$ and $\eta_Z(I_Y)$. Furthermore, it is obvious that

$$\eta_X(I_Y) = \eta_Z(I_Y) = \eta_Z(I_X).$$

The intensities of the components polarized parallel to X and Z, respectively, may be called

$$\xi = \xi_X(I_X) + \eta_X(I_Y) = \xi_X(I_X) + \eta_Z(I_X)$$

and

$$\eta = \eta_Z(I_Y) + \eta_Z(I_X) = 2\eta_Z(I_X).$$

The polarization is then given by

$$\bar{P}_0 = \frac{\xi - \eta}{\xi + \eta} = \frac{\xi_X(I_X) - \eta_Z(I_X)}{\xi_X(I_X) + 3\eta_Z(I_X)} \qquad \ldots\ldots(219).$$

If, on the other hand, we had measured the polarization (in zero field) of resonance radiation, excited by light of intensity I_X polarized in the X direction, we should have found a component polarized parallel to X, $\xi_X(I_X)$, and components circularly polarized about X, of which we see only the projection $\eta_Z(I_X)$. The polarization in this case is

$$P_0 = \frac{\xi_X(I_X) - \eta_Z(I_X)}{\xi_X(I_X) + \eta_Z(I_X)} \qquad \ldots\ldots(220).$$

From Eqs. (219) and (220) it follows that

$$\overline{P}_0 = \frac{P_0}{2 - P_0}, \quad \text{or} \quad P_0 = \frac{2\overline{P}_0}{\overline{P}_0 + 1} \quad \ldots\ldots(221).$$

In using the above formulas it is convenient to associate a plus sign with radiation partially polarized parallel to X and a minus sign with that polarized parallel to Z. In applying the formulas the signs of the polarization must be considered. For example, if $P_0 = +0.5$, then $\overline{P}_0 = +0.333$; or if $P_0 = -0.5$, $\overline{P}_0 = -0.20$.

INDEX